Anatomy of an Entangled Black Hole: Complexity is Information

First Edition

Dr. Robert Nieves

Library of Congress Control Number: 2023901480

ISBN: 9798374351248

© 2023 by Dr. Robert Nieves
All rights reserved

Anatomy of an Entangled Black Hole: Complexity is Information

This book is ideal for students, researchers, and readers, that seek an insightful understanding about black hole physics and quantum entanglement theory, written with the curious innovator in mind. Anatomy of an Entangled Black Hole probes the very limits of quantum entanglement and scientific possibility. Covering from the types and sizes of black holes to the latest research on the complexity of entangled black holes, it will take a fascinating scientific journey through the characteristics of space-time and into one of the most mysterious objects found in the universe. Written in a very readable style, from the attributes of the space-time around you to the latest research investigation about our universe. It conveys the latest concepts in black hole physics in a clear and understandable way. It is also filled with enlightening questions that are designed to motivate your imagination on the nature of black holes and spatiotemporal bridges.

There are three principal ideas in this book. The first idea is that complexity is information and quantum gravitation. The second idea is that two or more black holes may be entangled to form a black hole pair that may be traversable by information through a spatiotemporal bridge. The third idea is that highly energetic photons, or gamma rays, generate an outward pressure, producing the curvature of an outward counter-gravitational field.

In a number of chapters, a developmental overview is given of the key topics of what happened in black hole physics in the last three hundred years. It is a selection of topics on the theoretical physics of black holes that are suitable for understanding black hole types and sizes, black hole formation and evaporation, gravitational collapse, and gravitational emergence from entanglement to topics on thermodynamics of black holes, complexity, and Hawking radiation. In some chapters, spacetime is treated as an emergent field which diverges or converges inside or outside of a black hole, and black holes are discussed in asymptotically anti-de Sitter spacetime (AdS), for the relevance and motivation for these discussions originate from the AdS/CFT correspondence, $ER \equiv EPR$, and the emergence of complexity.

Some of the questions that are addressed in this book, including, but not limited to: What is entanglement? What is an entangled black hole? What does a black hole pair consist of? What is the event horizon of a black hole? What is the throat of an entangled black hole pair? What does the throat membrane consist of? What is a Planckian black hole? What goes on inside the event horizon of a black hole? What does complexity have to do with the saturation of a black hole? What does bound state energy or electron degeneracy have to do with counter gravitation? A provocative new book. Enlightening and engaging. Valuable insight into black hole physics! The author describes in detail these challenging subjects and how they impact entangled black holes.

Robert Nieves has a diversified professional experience in engineering, teaching, international business administration, and physics and cosmology research. Dr. Nieves holds a Bachelor of Science in Electrical Engineering from the Illinois Institute of Technology and an MBA and a DIBA from Nova Southeastern University in Florida.

Dedicated to my Father in heaven, Lord of Lords, to his son Jesus Christ, his mother Mary, and his Apostles, and to the Lord's Prophets and Believers, with all my Love.

CONTENTS

CHAPTER 1 — 1

An Overview of Black Holes.

1. What is a black hole?

    1.1. The formation of a black hole.

    1.2. The sizes and types of black holes.

2. Descriptions of black hole types.

    2.1. The Schwarzschild black hole.

    2.2 The Reissner-Northström black hole.

    2.3. The Kerr black hole.

    2.4. The quantum membrane of a black hole.

CHAPTER 2 — 8

The Attributes of Black Holes.

1. On the formation of a supermassive black hole.

    1.1. The radial coordinate of a Kerr-Newman black hole.

    1.2. The Schwarzschild factor during the formation of a singularity.

    1.3. The metric of a Kerr-Newman black hole.

2. Other attributes of black holes.

    2.1. An entangled black hole - white hole pair.

    2.2. On the existence of black holes.

3. On the creation of black holes and other celestial bodies in the universe.

CHAPTER 3   28

The Quantum Complexity of Space-Time Inside the Event Horizon of a Black Hole.

1. The emergence of quantum complexity.

    1.1. The measurement of all things.

    1.2. A Planckian black hole.

    1.3. The entanglement entropy toward a minimal area.

2. The monogamy of entanglement.

3. The holographic principle of the event horizon of a black hole.

4. The Complex Ginzburg-Landau Equation (CGLE).

5. Gravitation inside the event horizon of an entangled black hole pair.

CHAPTER 4   69

The Laws of Complexity.

1. The first law of quantum complexity for entangled black holes.

2. The second law of quantum complexity for entangled black holes.

3. The third law of quantum complexity for entangled black holes.

4. The fourth law of quantum complexity dispersion for entangled black holes.

5. The retarded and advanced wavelets of the arrow of time at a black hole.

6. The principle of equivalence of the event horizon of a black hole.

7. A quantum continuous equation for the decoherence and the evaporation of an entangled black hole pair.

8. About the quantum information of a free-falling object of mass, energy and space-time into a black hole.

   8.1. The Hawking radiation theory.

9. The Lambda CDM Model as the observed amount of photon sources in the universe.

CHAPTER 5   92

The Holographic Principle.

1. The $ER \equiv EPR$ conjecture.

2. The principles of entanglement for all there is inside the event horizons of black holes.

   2.1. Photon production from the spatiotemporal field.

CHAPTER 6   106

Will Our Universe Diverge and Then Converge like an Evaporating Black Hole?

1. The universe as a black hole - white hole pair.

   1.1. The converging universe.

   1.2. On the formation of a black hole.

   1.3. The point-like mass at the center of a collapsed star is known as a Schwarzschild singularity.

   1.4. So, how could a celestial body change its mass?

   1.5. How and when were the existing black holes detected?

1.6. How do we see or detect the invisible?

1.7. How could the gravitational field of a supermassive black hole be detected at a galactic distance?

2. Hawking's micro black holes.

2.1. Black hole collisions.

2.2. Did astronomers recently find the fastest spinning black hole in our universe?

2.3. The spatiotemporal curvature versus the gravitational potential for celestial bodies.

2.4. What is the minimum curvature in the spatiotemporal medium?

CHAPTER 7   133

The Invisible Dark Star.

1. The enigma of a black hole.

2. The astrophysics of a massive black hole.

2.1. Not even light can escape.

2.2. Is a dark star also a black hole?

2.3. Are the laws of Quantum Mechanics contradictory?

2.4. Are the correlations of Quantum Mechanics due to spacetime?

2.5. How could black holes be described today?

2.6. The collapsing dark star.

2.7. The formation of a theoretical white hole.

3. The tenses of time.

   3.1. The curvature from a triangle.

   3.2. The curvature from a hypercube.

   3.3. The Tolman-Oppenheimer-Volkoff counter-gravitational equation.

   3.4. Solving the Tolman-Oppenheimer-Volkoff equation for pressure or energy density.

   3.5. The Jacobian of spherical coordinates.

   3.6. The Jacobian spatiotemporal scale factor for a tesseract.

   3.7. The Jacobian for the volumetric transformation of spatiotemporal coordinates.

   3.8. Tic-Tac-Zero time at the event horizon.

   3.9. The spatiotemporal deformation potential of divergence or convergence.

REFERENCES   **207**

Chapter 1

An Overview of Black Holes.

§ 1. What is a black hole?

Black Holes are the most massive and probably the densest celestial objects in the observable universe. There are three main processes that produce black holes. Black holes act like one way street to a dead end, their gravitational field is so strong that not even light, or electromagnetic radiation, can escape once it crosses the event horizon of a black hole. A black hole, or dark star, is thought to be the kind of celestial object from which nothing can ever escape. The thermal property and the property of quantum gravitation are theorized to concurrently emerge within the event horizon of a black hole during saturation; more on that theory in the following chapters.

There are different types and sizes of black holes in the universe as we will discuss subsequently. Black holes are captivating objects where the effects of spatiotemporal curvature and gravitation are spectacular.

1.1. The formation of a black hole.

Currently, there are three main theoretical processes that form black holes. The first process is that a star, with about nine times the mass of "sol," our sun, may go supernova, which may entail the collapse of the star into a black hole. The strong nuclear force of a neutron star holds it back from collapsing as it stops the center of its atoms from collapsing. Nevertheless, a star with nine solar masses is so massive that its gravitation overcomes the strong nuclear force and completely collapses the atoms of the star into a singularity, or single point mass, or a ring, depending on black hole type or theory, with an enormous density.

A second process is the Hawking's quantum cosmological theory that trillions of black holes were produced in the Big Bang and some of those black holes still may exist today. A black hole with one solar mass may last longer than the present lifetime of the universe. However, black holes are theorized to emit radiation, so, they lose mass and eventually evaporate. A black hole with an approximate mass of 10^{12} Kg formed at the beginning of the Big Bang would be evaporating now. A third process for the formation of a black hole is when two neutron stars form a binary system, and they lose energy through gravitational radiation, slowly spiraling toward each other until they merge, typically producing a black hole.

It is theorized that black holes may be produced in different processes, but they all eventually evaporate the same way. The outgoing radiation may explain the process of black hole formation in its complexity. Two massive stars may orbit each other very fast as a binary system, producing gravitational waves, ripples in the fabric of space-time.

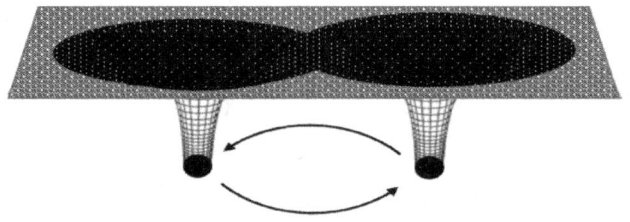

Figure 1. An Illustration of a Binary Dark Star System.

1.2. The sizes and types of black holes.

There are at least three properties by which a black hole may be classified: Mass, Magnetic Field, and Spin.

The observed or theorized types of mass are: stellar mass, intermediate mass, and supermassive. The black holes with stellar masses are about ten to a hundred times the mass of sol. Black holes with intermediate masses are hypothesized but they are still to be found, and their formation mechanism need to be theorized. The center, or core, of every galaxy is theorized to have a supermassive black hole. The supermassive black holes may have millions or billions of solar masses. The holographic principle may describe the spatiotemporal medium inside black holes in terms of a theory on the outer boundary, or outer event horizon.

There exists four exact solutions of Einstein's Field Equations describing black hole solutions with or without charge and angular momentum:

- The Schwarzschild solution (1917) has only mass M; it is static, and spherically symmetric.
- The Reissner-Nordström solution (1918), it is static, and spherically symmetric; it depends on mass M and electric charge Q.
- The Kerr solution (1963), it is stationary and axisymmetric; it depends on mass and angular momentum.
- The Kerr-Newman solution (1965), it is stationary and axisymmetric, it depends on all three parameters M, J, and Q.

The black holes that are theorized to lack or to have a magnetic field or spin are:

A Schwarzschild black hole has no magnetic field and no spin. It is the simplest type of black hole to be theorized. A Reissner-Nordström black hole has a magnetic field but no spin. A black hole that does not spin may be perfectly spherical. A Kerr black hole has both spin and a magnetic field, so it has spin angular momentum. The following table shows four black hole types that include charge "Q" and spin angular momentum "J" for a black hole type.

	Non-Rotating ($J = 0$)	Rotating ($J \neq 0$)
Uncharged ($Q = 0$)	Schwarzschild	Kerr
Charged ($Q \neq 0$)	Reissner-Nordström	Kerr-Newman

Table 1. Four related black hole solutions are summarized in this table.

§ 2. Descriptions of black hole types.

2.1. The Schwarzschild black hole.

The Schwarzschild black hole is the simplest type of black hole, it has a boundary called an event horizon, and a singularity. The singularity is a single point mass left from the collapsed star with finite mass and enormous density. It is theorized that inside the event horizon the radial spatial velocity is greater than the speed of light, so once something crosses the boundary of the event horizon, nothing can escape the black hole.

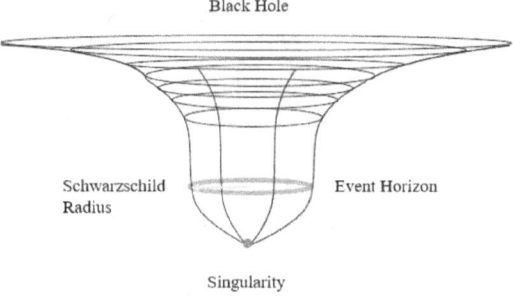

Figure 2. An Illustration of a Schwarzschild Black Hole.

2.2. The Reissner-Nordström black hole.

The Reissner-Nordström black hole has two even horizons, a singularity, and is electrically charged. At the outer event horizon boundary, space becomes time and time becomes space. The singularity becomes a single point mass in the temporal future, instead of in space. The inner event horizon reverts time into space and space into time, as it would be in the external spatiotemporal medium.

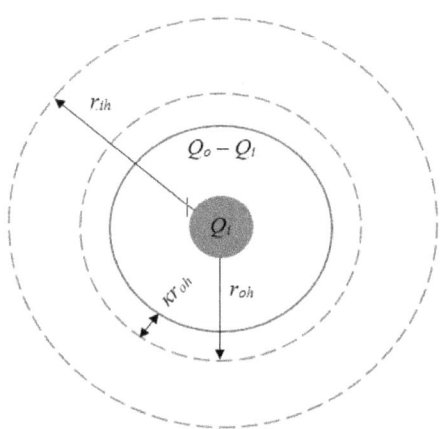

Figure 3. An Illustration of a Reissner-Nordström Black Hole.

If the two event horizons are crossed, space and time exchange roles twice, so, inside the sphere of the inner horizon, also known as the Cauchy horizon, space and time revert to their usual roles. Hence, an infalling object may avoid the temporal singularity. If the black hole charge is very high, the two event horizons may disappear and the singularity becomes a naked singularity. Thus, in the Reissner–Nordström black hole, the singularity inside is time-like and the inner horizon is an unstable Cauchy horizon.

2.3. The Kerr black hole.

A Kerr black hole has an ergosphere, which is an elliptical region outside the event horizon. The ergosphere may be navigated by a spacecraft in a last stable orbit around the black hole. Outside the ergosphere, the spacecraft could navigate unobstructed. Inside the ergosphere, the spatiotemporal medium is curved and drags the spatiotemporal frame inward. A Kerr-Newman black hole is a Kerr black hole which is both rotating and electrically charged.

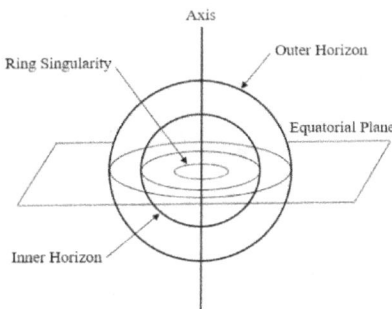

Figure 4. An Illustration of a Kerr Black Hole.

The faster the Kerr black hole spins, the larger the inner event horizon becomes, while the outer event horizon stays the same proportion. Nevertheless, when the rotational energy equals the energy of the mass, the outer and inner horizons become alike. If the rotational energy were to be greater than the energy of the mass, the horizons would disappear leaving behind a naked singularity. A distinctive characteristic of a Kerr black hole is its ring singularity. The rotation of the Kerr black hole spins its singularity into an infinitesimal ring, where it is theorized that nothing can enter the ring singularity unless it approaches the singularity along the side of the ring. Any other angle of approach repels the incoming mass or matter with a counter gravitational force.

2.4. The quantum membrane of a black hole.

A quantum membrane has been identified by a group of international physicists when they examined the structure of a gravitational wave emitted by two merging black holes. The phenomenon of a quantum membrane has been sought after for a long time.

The membrane is also known as a stretched event horizon, that exists precisely outside the event horizon of a black hole, which affects the behavior of the gravitational and non-gravitational fields near a black hole. The membrane had been predicted by some of the quantum gravity theories. Quantum gravity theories attempt to find how to describe gravitation in terms of quantum mechanics. (Abedi, 2021)

The theory of everything has been sought after in physics, first through the Grand Unified General Theory of Relativity and Quantum Mechanics, for more than a hundred years, including several quantum gravity theories have been put forward. The Grand Unified Theory should clarify and resolve long standing questions in modern physics with regards to the Big Bang process at the beginning of our universe and the problem of a singularity with an infinite density at the center of black holes.

The realization of an experiment that could verify these matter is very complex due to the quantum scale of the energy scale for quantum gravitational effects. The effects of quantum gravitation become significant at many orders of magnitudes above the energy level of particles that can be obtained in a particle collider. It is crucial to seek another way to study quantum gravitation.

The separation between the quantum membrane and the event horizon of a black hole is close to the Planck length "ℓ_p," an applicable length to the scale of quantum gravitation, a spatial distance that is less than the diameter of an atom. It is theorized that at the Planckian scale, spatiotemporal medium is a medium that emerges from the interference of spatiotemporal wavelets that "A Dynamic Theory of Space-Time" can describe as it explicates what happens at the quantum scale or above. (Nieves, 2020 and 2021)

The existence of the quantum membrane allows a gravitational echo to occur in the spatiotemporal medium of the black hole. The event horizon of the black hole surrounds its geometrical volume, where the gravitational field of the black hole sets up a perimeter of no return back into outside space-time, if anything falls through the event horizon it cannot escape, not even gravitational waves or light.

Chapter 2

The Attributes of Black Holes.

§ 1. On the formation of a supermassive black hole.

Let us imagine that space is able to expand faster than light and that the physical laws of the universe remain the same to allow non-homogeneous and anisotropic space to expand at several times the speed of light c, through several consecutive light barriers such as c, $2c$, $3c$, $4c$, $5c$, etc; such as in the case of the formation of a charged rotating supermassive Kerr black hole.

Mass and time condense, and length expands at every light barrier reached, seemingly making a very infinitesimally small relativistic mass m' appear to disappear, when the contracted mass minimizes its dimensions from its previous volume before arriving at the light barrier. At the same instant, relativistic time is highly dilated, so when the mass condenses at the light barrier, time is proportionally condensed from its previous temporal magnitude. However, time speeds up as it condenses at each light barrier. As time condenses, the relativistic mass travels faster, and time speeds up as length expands. Furthermore, the gravity field of the mass decreases proportionately, as mass and time contract, during every light barrier event.

The electromagnetic forces of the unified electroweak interaction of the mass and charges become stronger than the gravitational field of the mass during every light barrier event. As atomic structures, charge distribution and energy collapse into a tighter, more compacted mass distribution. The increasing pressures and temperatures allow the positively charged nuclei of matter to combine. As the mass coheres, charges condense under the immense tidal forces of the collapsing body, creating an increasingly cohesive charge distribution about the geometry of the forming singularity, springing very strong electric fields out of the condensed singularity as charges focally direct their electric field lines radially outward.

As the density of the mass increases during the collapse of the celestial body, the surface area of its space-time-mass boundary gets smaller and the force of space acting on the mass has a decreasing surface area to act upon. Thus, it is postulated that as the mass compacts to a singularity, the pressure of space on the infinitesimal mass surface exerts a gravitational field, until the mass of the singularity collapses into the temporal dimension of space-time, outside of local space, in what becomes the outer event horizon of the Kerr black hole. Thus, the space-time flowing inward lengthens, and space pressure lessens and exerts a lesser potential gravitational effect on mass as space expands inward. Hence, as the mass speeds up through light barriers, its gravity field weakens and its electromagnetic field strengthens, as the mass density increases. The strength of the electromagnetic field forces of the unified electroweak interaction of the mass and charges acts on the area of the outer event horizon of the Kerr black hole preserving its geometry and boundary, creating an effective demarcation between space and time. This action sustains the concentrated electromagnetic field strength of the singularity and also establishes a stable gravitational field for the emerging black hole.

As the singularity forms, the weak interaction can convert nucleons of one kind to the other, emitting an electron or a positron in the process through beta radioactivity. The strong interaction grows stronger as it binds nucleons together and the nuclei combine under the extreme temperatures and pressures of the extremely dense mass. As the mass continues to collapse further within the temporal dimension of the outer event horizon, a spatial distance develops and its boundary is the inner event horizon of the Kerr black hole. As the singularity reaches a stable existence as a point mass or an infinitesimal mass, it exists within the inner event horizon space with infinitesimal nonzero dimensions. Thus, it is postulated that the singularity has attributes of very condensed mass, spatial dimensions and it exists in an infinitesimal localized space-time bubble within the black hole. However, the singularity exists in localized space outside of the universal space-time beyond the ergosphere of the black hole. This condition decouples the potential gravitational effects of the singularity's mass within the inner event horizon from the outside universal space-time.

As the geometry of the black hole is sustained by the concentrated electromagnetic field, a gravitational field is exerted to and beyond the outer event horizon and ergosphere of the supermassive black hole, as if the collapsed celestial body were still there in physical form and with its corresponding mass. Hence, according to the above theoretical gravitational model, the final gravitational field of the supermassive black hole results from the space-time pressure differential sustained by the concentrated electromagnetic field of the unified electroweak interaction of the mass and charges at the space-time boundary of the outer event horizon and not necessarily as a gravitational end result of the finally condensed mass of the singularity itself.

The gravitational field of a Kerr black hole is enough to draw matter in and keep the matter spinning in a stable accretion disk. Material and gas accumulate in an accretion disk around a Kerr black hole. The spinning material and gas generate their own magnetic fields, and these fields power winds of charged particles that blow away from the black hole. The winds transfer angular momentum from the inner regions of the disk outward. This slows down some of the spinning material and gas, allowing the material and gas to fall onto the black hole. But before matter in the accretion disk can take the final plunge into the ergosphere and then into the outer event horizon of the black hole, the matter must lose some of its rotation speed or angular momentum. If the angular momentum from the disk were not dissipated away, the material and gas in the accretion disk would circle the black hole a very long time in a stable orbit, like planets circling a star.

It is postulated that a torsional Alfvén wave is generated by the rotational dragging of space near a Kerr black hole. The wave transports energy outward along the magnetic field lines, causing the total energy of the plasma near the black hole to decrease to negative values. The magnetic field also causes turbulence and friction to build up within the disk. The friction heats up the gas to millions of degrees, causing the plasma to glow brilliantly in the ultraviolet and X-ray bands. When this negative energy plasma enters the outer event horizon, the rotational energy of the black hole decreases. Through this process, the energy of the spinning black hole is extracted magnetically. The magnetic coupling process can transfer energy and angular momentum from a rotating Kerr black hole to its surrounding disk. Recently, it has been observed through powerful special telescopes that a black hole's powerful magnetic fields create turbulence in surrounding matter that help drive matter inward to be accreted. It is postulated that tightly coiled magnetic fields close to energetic supermassive black holes expel narrow jets of plasma into space. This mechanism transfers heat to gases attracted to and near the black hole to effectively control the growth of the largest galaxies.

1.1. The radial coordinate of a Kerr-Newman black hole.

Furthermore, as a Kerr-Newman black hole forms as previously proposed, the very condensed mass of the celestial body collapses beyond the radial distance "s" of space and into the radial distance "τ" of time. Thus, as length extends, time contracts, and the radius of the ring can be measured in terms of its temporal distance. In other words, the ring of the black hole has moved within its temporal bubble as time condenses, outside of external space-time and within the outer event horizon. This physical condition within the outer event horizon allows space-time to flow and curve inward as a river in the direction of the ring into the inner space-time of the supermassive Kerr-Newman black hole. Hence, from the following integral expression for the spatial and temporal radial coordinates of the collapsing mass of the celestial body, we find

$$r = \int_{s_1}^{s_2} \partial s_r + c \int_{t_1}^{t_2} \partial t_r \quad \text{where } s_2 > s_1 \text{ and } t_2 < t_1 \tag{1.1}$$

$$r = s_r \Big|_0^S + ct_r \Big|_{t_1}^{t_2} \tag{1.2}$$

$$r = s + c(t_2 - t_1) \tag{1.3}$$

Let the temporal interval $(t_2 - t_1)$ be equal to τ to find the radial distance "r," as the spatial distance "s" extends and the mass collapses into a ring singularity, where it is theorized that the internal electromagnetic field and internal gravitational field are repulsive, or negative, from the perspective of the external gravitational field outside of the outer even horizon.

$$r = \lim_{m \to 0} [s + c(t_2 - t_1)] = c\tau \qquad (1.4)$$

Hence, we specify from this physical model that the radius "r" of the ring is no longer spatial, but it is temporal. The value "$c\tau$" is the distance of the radius "r" in the spatiotemporal dimension, within the outer event horizon of a Kerr-Newman black hole.

1.2. The Schwarzschild factor during the formation of a singularity.

Let us consider the Schwarzschild's geometry of a black hole, where $r_s = 2GM/c^2$, about the event horizon with $r > r_s$, described by the following metric:

$$ds^2 = -\left(1 - \frac{r_s}{r}\right)dt^2 + \frac{dr^2}{\left(1 - \frac{r_s}{r}\right)} + r^2 d\theta^2 + r^2 \sin^2\theta d\phi^2 \qquad (1.5)$$

where we can express the Schwarzschild factor as

$$1 - \frac{r_s}{r} = 1 - \frac{2GM}{rc^2} \qquad (1.6)$$

As we reconsider the radial coordinate r, with space flowing radially inwards at the Newtonian infalling speed of $v^2 = 2GM/r$, in General Relativity, as "v" reaches the speed of light "c", we have

$$\lim_{v \to c}\left(1 - \frac{2GM}{rc^2}\right) = \lim_{v \to c}\left(1 - \frac{v^2}{c^2}\right) = \lim_{v \to c}\left(1 - \frac{c^2}{c^2}\right) = 0 \qquad (1.7)$$

Therefore, as spatial distance within the black hole extends, it is theorized that the Schwarzschild factor approaches zero as predicted by General Relativity, within the internal region of the theoretical supermassive black hole. The infalling speed "*v*" passes the speed of light "*c*" at the event horizon of the black hole. As $r < r_s$, the speed of contracting of the spatiotemporal wave increases, as space extends and time condenses. Thus, under these proposed conditions Einstein's law regarding the speed of light still holds because the speed of light applies to the speed of objects moving in space-time as measured locally with respect to an inertial frame of reference. In this model within the event horizon, it is space-time itself that moves superluminally inward; Hence, theoretically, the General Theory of Relativity prevails.

1.3. The metric of a Kerr-Newman black hole.

The Kerr–Newman metric describes the geometry of space-time in the vicinity of a rotating black hole of mass *M*, spin α, with a charge *Q* in standard spherical coordinates.

$$c^2 d\tau^2 = -\left[\frac{dr^2}{\Delta} + d\theta^2\right]\rho^2 + \left[cdt - \alpha\sin^2\theta d\phi\right]^2\frac{\Delta}{\rho^2} - \left[(r^2+\alpha^2)d\phi - \alpha cdt\right]^2\frac{\sin^2\theta}{\rho^2} \quad (1.8)$$

$$\alpha = \frac{J}{Mc} = \frac{a}{c} \quad (1.9)$$

$$\rho^2 = r^2 + \alpha^2\cos^2\theta = r^2 + \frac{a^2\cos^2\theta}{c^2} \quad (1.10)$$

$$\Delta = r^2 - r_s r + \alpha^2 + r_Q^2 \quad (1.11)$$

$$r_Q^2 = \frac{Q^2 G}{4\pi\varepsilon_0 c^4} \quad (1.12)$$

Hence, the current capacity at the radius r_Q corresponding to the charge *Q* of the Kerr-Newman black hole with a mass *M* can be expressed as

$$I_Q^2 = \frac{4\pi\varepsilon_0 c^4 r_Q^2 f_Q^2}{G} = Q^2 \left(\frac{\ddot{m}}{M}\right) \tag{1.13}$$

$$\left(\frac{I_Q}{Q}\right)^2 = \frac{\ddot{m}}{M} \tag{1.14}$$

where $\dot{m}$ is the matter accretion velocity rate, and $\ddot{m}$ is the matter accretion acceleration rate of the Kerr-Newman black hole at the radius r_Q.

$$\dot{m} = \frac{I_Q M}{Q} \tag{1.15}$$

$$\ddot{m} = \frac{I_Q^2 M}{Q^2} \tag{1.16}$$

Let us express the Kerr-Newman metric in terms of the radial coordinate "r," the angular momentum "a" per unit of mass, using the space-time geometry in the vicinity of the rotating Kerr-Newman black hole.

$$c^2 d\tau^2 = -\left[\frac{dr^2}{r^2\left(1-\frac{r_s}{r}\right)+\frac{a^2}{c^2}+r_Q^2} + d\theta^2\right]\left[r^2 + \frac{a^2 \cos^2\theta}{c^2}\right] + \left[cdt - \frac{a\sin^2\theta\, d\phi}{c}\right]^2 \frac{r^2\left(1-\frac{r_s}{r}\right)+\frac{a^2}{c^2}+r_Q^2}{r^2 + \frac{a^2\cos^2\theta}{c^2}}$$

$$-\left[\left(r^2+\frac{a^2}{c^2}\right)d\phi - adt\right]^2 \frac{\sin^2\theta}{r^2 + \frac{a^2\cos^2\theta}{c^2}} \tag{1.17}$$

As the speed of space-time approaches the speed of light c, the Schwarzschild factor $(1-r_s/r)$ approaches zero as predicted by the

General Theory of Relativity and the Kerr-Newman metric for $r \leq r_S$, where $r^2 = c^2 t^2$, $r_Q = c^2 t_Q^2$ and $dr^2 = c^2 dt^2$, can be expressed as

$$c^2 d\tau^2 = -\left[\frac{c^2 dt^2}{\frac{a^2}{c^2} + c^2 t_Q^2} + d\theta^2\right]\left[c^2 t^2 + \frac{a^2 \cos^2 \theta}{c^2}\right] + \left[cdt - \frac{a\sin^2\theta\, d\phi}{c}\right]^2 \left[\frac{\frac{a^2}{c^2} + c^2 t_Q^2}{c^2 t^2 + \frac{a^2 \cos^2\theta}{c^2}}\right]$$

$$-\left[\left(c^2 t^2 + \frac{a^2}{c^2}\right)d\phi - adt\right]^2 \left[\frac{\sin^2\theta}{c^2 t^2 + \frac{a^2 \cos^2\theta}{c^2}}\right] \quad (1.18)$$

§ 2. Other attributes of black holes.

It is interesting to note that the event horizons of black holes are not physical, but only calculated boundaries, for the spatiotemporal characteristics of the medium, produced by the type of singularity and its field effects. It is worthy to note that a Reissner-Nordström or a Kerr-Newman black hole is gravitationally repulsive at its core. Indeed, the singularity is infinitely gravitationally repulsive.

In a rotating black hole or in a black hole with a magnetic field, an accretion disk may form due to the mechanical forces that exist near the black hole. An accretion disk forms from the matter that is drawn to the black hole. The matter and mass in accretion disks are gradually drawn into the black hole, and as the matter or mass draws closer, its kinetic energy increases. The very high temperature of an accretion disk may go up due to internal friction to billions of Kelvin, while high-speed jets of matter, gamma rays and other radiation may be emitted from the poles of the black hole into the spatiotemporal medium. All the observable radiation, or the velocity of the matter using the doppler effect, may be utilized to calculate the approximate mass of the black hole.

For a nonrotational black hole like a Schwarzschild black hole, the matter or mass is drawn in equally from all around, forming a cloud of radial accretion instead of a disk.

2.1. An entangled black hole - white hole pair.

Quantum entanglement builds bridges! It is theorized that an entangled black hole - white hole pair has a spatiotemporal bridge, which extends through contracted space between two or more present spatial regions, that may be traversable by information. Information is contrast between objects, which is the state of being strikingly distinct from something else placed close together for careful comparison, or in juxtaposition. A white hole is the conjugate of the black hole. The black hole is seen as the entry point of the spatiotemporal bridge and the white hole is seen as an exit point. So, it is theorized that the white hole ejects matter and energy out. Furthermore, it is theorized that an object falling into a traversable spatiotemporal bridge may travel on the advanced wavelet toward the past, but as it traverses the throat through the middle membrane, the object travels on the retarded wavelet toward the future, and may exit the bridge at, or nearly at, the same time as it entered. More on the physics of entangled black hole - white hole pair in the following chapters.

2.2. On the existence of black holes.

Black holes have been theorized and identified but not directly observed with a telescope. Black holes have been observed indirectly by observing how the matter around the black hole moves. Black holes are found in the center of galaxies.

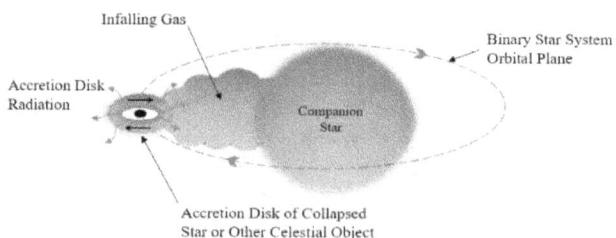

Figure 1. A Black Hole Accreting Matter from a Companion Star.

It is hypothesized that black holes continue to converge after saturation as long as the outside space-time is diverging. If the outside space-time were converging, the black hole would diverge. In the case of a micro black hole when the outside space-time diverges long enough, the black hole may evaporate.

Under the theories of special relativity and quantum mechanics for an accelerated massless observer, the energy of the observer through space-time would become a boost generator. If the accelerated observer were to continue through Euclidean space, the boost would become a rotation. It is hypothesized that if the speed of the observer were to approach the speed of light, the rotation would continue to redirect the observer toward the opposite direction, or the direction of the advanced spatiotemporal wave, until the motion is superluminal from the perspective of the retarded wave, as previously stated in the Dynamic Theory of Space-Time.

An inflationary period of expansion with nearly constant acceleration would produce a nearly isotropic and homogeneous universe of small quantum fluctuations with corresponding temperatures. It is theorized that when a virtual particle falls through the outer event horizon of a Kerr-Newman black hole it could spread over all the degrees of freedom. However, if its anti-particle partner stays out of the black hole, it may have a set of opposite quantum numbers to its particle partner. Hence, inaccessible information may not need to be extracted but inferred. If it were possible that a particle of a known virtual pair were purposely thrown into a black hole, the remaining anti-particle may provide its quantum numbers.

§ 3. On the creation of black holes and other celestial bodies in the universe.

Each universal law is an eternal and inherent nature of complex existence or reality that ultimately underlies a right behavior, wisdom, and behavioral norm for mass, matter, energy, waves, or lifeform. devised by the creator of all there is. Each law is an instance of truth, or of reality, that is an aspect, or physical form, of the spatiotemporal volume, where everything comes from the ideas of faith. It may come to past that there is such a thing as a love correspondence, when physics finds that time is the emergence of more space, and that time emerges from the will of a real creator for the love of his creation. It may all come down to "$time \equiv |love|^2 e^{love}$" for all there is in terms of complex physics, since there are different kinds of love. Faith is freedom and joy. Love is the eternal prime mover for the best insights and ideas which come from the universal creator at the perfect time. Lo and behold, there was light at the beginning of creation, as well as in the birth of complex physics. *Physics cannot explicate creation by referring only to physics, it takes divine design and faith to provide a conceptualization of a complex existence or reality that includes the entanglement of physics and spirituality.* So, instead of peering at the vacant murk hiding from light, why not gaze at the brightness that shines on our eyes. For the energy of eternal truth and joy is to life what the constant brightness is to light. Praise be every day and night, to the creator of the fountain of life, at least seventy-seven times. Take care of all that is possible in life, or Complex Physics, that the Lord will take care of the impossible. If you believe in the Lord, bend your knees in prayer, for all that is impossible in your life, or in Complex Physics, will become possible.

Let us posit or imagine the complex existence of our universe consisting of a real part (physical reality) and a spiritual part (nonphysical reality), if only for the faithful at heart, or for a scientist that can treat both sides of complex existence on an equal footing. The physical aspect has a structure of atoms, molecules, and forms, where form precedes thought, a brain, a body, physical senses, and a soul; all existing physical aspects. A sense is a biological system used by a lifeform for sensation, the process that gathers information about the physical environment through the detection of physical stimuli. The physical senses include, but may not be limited to, sight, touch, hearing, taste, smell, orientation, and thought, Let us define the soul as a spherical volume of energy, an energy field, with a central nucleus of point-like energy within an energy boundary or membrane that may be visible in the infrared spectrum at a distinct frequency. The insights, dreams, and imprints of a soul may occur as a scale translation to higher frequency volumes of complex existence. The soul is a reusable container or depository of, but not limited to, a mind, memories, feelings, knowledge, and life experiences through the physical senses, for a lifeform, that is entangled to the body through a micro spatiotemporal bridge, which is a nonlocal connection according to General Relativity.

Entanglement is Complexity. A nonlocal connection between physical volumes of distinct frequencies. So, the mind of a lifeform is part of its soul. The soul has mass and it may travel with the body, in its frame of reference, typically at a speed much lower than "c" or it may vary its frequency. Physical reality and spiritual reality have very distinct complexities. *The complexity of physical reality is theorized to emerge from the quantum divergence and convergence of the spatiotemporal medium while the complexity of the spiritual reality is theorized to emerge from thoughts or consciousness. Therefore, it is reasonable to hypothesize that physical reality may exist within spiritual consciousness.*

On the spiritual part or the nonphysical part, the spiritual medium may resemble a cytoplasmic medium. The cytoplasmic medium is a luminous gel-like fluid or solution, or a thick luminous semifluid or condensate at very high frequency, where thought precedes form. It is the luminous medium for temporary or permanent form creation and action; the medium that entangles the spirit to the complex existence of its collective. A luminous substance that may be manifested in the physical spatiotemporal medium as a shape or holograph of a non-solid volume. The spirit of a lifeform may shape its form, shapeshift to other forms, or entangle to the form of a spiritual collective medium, through thought. The spirit is entangled to the soul through a quantum spatiotemporal bridge of entanglement which is a nonlocal connection according to quantum mechanics.

Entanglement may be the nonlocal bridge process between the physical part and the spiritual part of complex existence. The spirit has consciousness that may accumulate a mind, memories, knowledge, imprints, and life experiences through the physical senses, of the soul. The spirit is massless, so it may travel, by thought, at relativistic speeds, or faster than light backwards in time, or vary its frequency, in the physical medium. The spirit may be disconnected from the soul if the behavior and course of the soul becomes unacceptable and irreconcilable, at which time the entanglement is terminated; the soul, its mind and attributes, at its reference frame, may become independent of the consciousness of the spirit, in which case it may be regarded stray from the perspective of the spirit and its true purpose.

How could the creator of our universe stand outside of time?

Black holes can rotate at nearly the speed of light. It is theorized that at the event horizon of a black hole time stands almost still. So, how is this possible? The Einstein-Cartan Theory modifies gravitation in the General Theory of Relativity, allowing space-time to have torsion, in addition to curvature, and relating torsion to the density of intrinsic angular momentum. which vanishes as soon as vacuum regions are considered. Torsion can also be related mathematically to spin.

The concept of an affine connection found an independent meaning, from the Levi-Civita connection of GR, during the period of 1918 through 1924 due to the work of the eminent physicists Elie Cartan and Hermann Weyl. There is no relation between the metric and the affine connection, where the metric is responsible to describe the causal structure, whereas the affine connection deals with the geodesic structure. (Cartan, 1923-1925, and Weyl, 1923)

In a Riemann-Cartan geometry, contrary to a Riemannian geometry, the affine connection is independent of the spatiotemporal metric of a linear trajectory, while the Riemannian geometry is derived from the linear or radial spatiotemporal metric as the Levi-Civita connection. The torsion, or the antisymmetric part, of the connection is zero for a Levi-Civita connection. Hence, the difference between the two geometries is referred as the spatiotemporal contorsion. The contorsion may be expressed linearly in terms of the torsion, then it becomes possible to directly translate the Einstein-Hilbert action into a Riemann-Cartan geometry, which is known as the constraint Palatini action or variation.

Therefore, the Einstein-Cartan Field Equations (ECFEs) may be expressed in terms of an affine connection, instead of a Levi-Civita connection. There would be terms in the ECFEs representing the contorsion that are additional to the EFEs that may be obtained from the Palatini formulation, and other terms for the additional torsion equations coupling to the intrinsic angular momentum, or spin, of matter, in a similar manner in which the affine connection is coupled to the momentum and energy of matter.

A geometric object, such as a spherical black hole, may be used to describe an affine connection on a smooth manifold to be connected to "nearby" tangent spaces; so, the affine connection defines a way to connect tangent vector fields to be differentiated as if they were functions on a manifold with values in a fixed vector space. There may be a tensor field, known as the torsion tensor, associated to every affine connection.

The tensors $T^r_{\theta\varphi}$ and $R^r_{\theta\varphi\phi}$ are, respectively, the torsion tensor and the curvature tensor of the affine connection on a smooth manifold M. The system $\Gamma^r_{\theta\varphi}$ is called the object of the affine connection.

$$T^r_{\theta\varphi} = \Gamma^r_{\theta\varphi} - \Gamma^r_{\phi\varphi} \tag{3.1}$$

$$R^r_{\theta\varphi\phi} = \frac{\partial \Gamma^r_{\theta\phi}}{\partial x^\varphi} - \frac{\partial \Gamma^r_{\theta\varphi}}{\partial x^\phi} + \Gamma^r_{\lambda\varphi}\Gamma^\lambda_{\theta\phi} - \Gamma^r_{\lambda\phi}\Gamma^\lambda_{\theta\varphi} \tag{3.2}$$

$$\omega^r = \Gamma^r_{\varphi} ds^\varphi \quad \text{where } \det\left|\Gamma^r_{\varphi}\right| \neq 0 \tag{3.3}$$

$$\omega^r_{\theta} = \Gamma^r_{\theta\varphi}\omega^\varphi \tag{3.4}$$

$$\text{if } r \equiv \theta, \; \Gamma^r_{r\varphi} = \frac{1}{2|g|}\partial_\varphi |g| \tag{3.5}$$

The geodesic lines may be defined by the system with respect to a local parallel coordinate system as follows:

$$\frac{d^2 x^r}{dt^2} + \Gamma^r_{\theta\varphi}\frac{dx^\theta}{dt}\frac{dx^\varphi}{dt} = 0 \tag{3.6}$$

Metric-Affine Teleparallelism can examine the nature of distant parallel lines in a geometric framework, to explore how both the curvature, torsion, and contorsion of space-time affects the motion of mass, matter, or energy. The affine connection represents a mathematical expression that describes how parallel lines are transported through spacetime along a curved manifold. Metric-Affine Teleparallelism proposes an independent network of connections between each point in spacetime that supplements the standard relationships of the General Theory of Relativity.

Let us theorize that Metric-Affine Teleparallelism may be achieved to connect two nearby tangent spatiotemporal smooth manifolds of ideal equilateral triangles. An equilateral triangle is a triangle in which all three sides are the same length. In Euclidean geometry, an equilateral triangle is also equiangular, since each of the three internal angles "φ_n" is also congruent to the other two and each is 60°. The orthocenter of the equilateral triangle is the point $P(r,\theta,\varphi)$, not regarded as a point of origin, where all three altitudes of the triangle intersect in a spherical coordinate system. The measurement in angles, or radians, preserves an important aspect of time, its absolute attribute, which becomes associated with metric-affine teleparallelism. Let us consider that a tetrad, a torsion spinor, or a phasor, may be used as an altitude with a counter rotating frequency "ω_n^r" and a magnitude equal to the Planck length "ℓ_p" or "ct_p" to represent the propagation of a photon from the orthocenter to each corner of the triangle. This photon may represent the behavior of the electromagnetic field across the spatiotemporal bridge of entanglement during Metric-Affine Teleparallelism. Hence, it is reasonable to state that Metric-Affine Teleparallelism, or teleparallel gravity, through entanglement may portray a space-time characterized by a potentially curved, torsional, and contorted nonlinear affine connection in conjunction with a tensor field independent of the spatiotemporal metric or line element, both defined in terms of a dynamical tetrad field.

The formalism of a tetrad "$e_a = e_a{}^\mu \partial_\mu$", where $a = 1,…, n$, that together span the n-dimensional tangent bundle at each point in the space-time manifold "M", is a method to the General Theory of Relativity that generalizes the choice of basis for the tangent bundle from a coordinate basis to the less restrictive choice of a local basis that may be defined as a set of four, or six, linearly independent vector fields.

A spinor transforms to its negative when the space-time is

continuously rotated through a complete turn from 0° to 360°, as it may happen in the throat of a black hole - white hole pair. Spinors are essential to describe the intrinsic angular momentum of particles. (Cartan, 1913) On the other hand, all three phasors may be teletransported along with the triangle to a nearby point to measure curvature, torsion, and contorsion, according to the ECFEs.

If there is curvature at the end of the teletransportation, all angles inside the triangle will be less or more than 180°; if there is torsion, the altitude phasors will have a different phase angle; if there is contorsion, the length, or magnitude, of any of the three altitude phasors would be less or more than the Planck length.

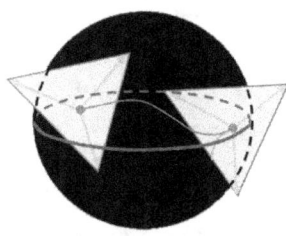

Figure 2. An Illustration of an Affine Connection on a Black Hole.

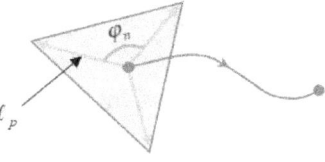

Figure 3. An Equilateral Triangle as a Spatiotemporal Smooth Manifold.

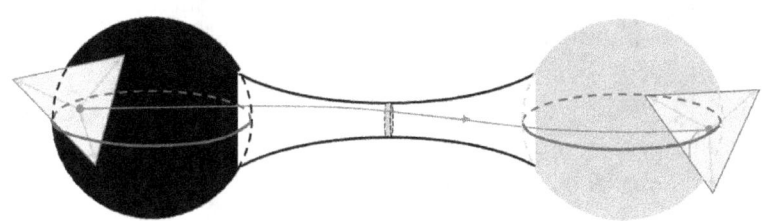

Figure 4. An Illustration of an Affine Connection across a Black Hole - White Hole Pair.

Consequently, let us imagine that all three phasors, tetrads, or torsion spinors, will be teletransported along with the triangle to a very distant point to measure curvature, torsion, and/or contorsion, according to the ECFEs, with the trajectory of the teletransportation going through the throat of a black hole - white hole pair, an affine connection, to a smooth manifold of the white hole to be connected to tangent subspaces. If parallel transport preserves torsion, the torsion vectors that are associated with the two parallel triangular surface elements are themselves parallel and the spatiotemporal medium is largely homogeneous.

It is important to point out that two phasors or two vectors whose infinitely close translating start points are parallel can be inferred from each other by the infinitesimal translation that brings their two translating points into conjunction. If there is no curvature, the two vectors can be immediately seen whether they are parallel or not, since the parallel transport of vectors becomes independent of the path.

Curvature governs how the tangent space moves on a curve on a manifold, torsion commands how the tangent space twists around a curve when we parallel transport two vectors along each other, non-metrical encrypts the difference in the altitude of a vector or phasor when they move along a curve. When parallel transporting a vector forward, the vector rate of change is completely normal to the surface of the given path. The covariant derivative "$\nabla_{\vec{d}} \vec{v}$" is the rate of change of a vector field in a specific direction "$\vec{d}$" with the normal component "$\vec{n}$" subtracted. If the covariant derivative is equal to zero, the vector field has been parallel transported.

Aside, an affine subspace, or hypoplane, of dimension $(n-1)$, in an affine space of dimension "n", is an affine hyperplane that may be inside the event horizon of the black hole and/or throat. It is interesting to note that the affine connection may serve as a lemma to connect tangent spatiotemporal smooth manifolds of ideal volumes through the throat between an entangled black hole - white hole pair within our universe, across space or time, or as an interdimensional affine connection, between universes, in an interdimensional metric-affine teleparallelism, under the principle of entanglement of the ER = EPR conjecture within interconnected spatiotemporal media.

The torsion in the ECFE becomes a variable in the principle of stationary action that couples to a curved spacetime spin tensor. These extra equations express the torsion linearly in terms of a spin tensor that is associated with the source of matter, which requires that the torsion would typically not be zero inside the source of matter.

The spin tensor $\sigma_{r\theta}{}^{\phi}$ represents the rotational motion of a particle in the spatiotemporal medium, which self-interacts nonlinearly with an effective spin-spin, in the internal spatiotemporal medium, or plenum, of matter.

The Einstein-Cartan Field Equations may be denoted as

$$R_{\mu\nu} - \frac{R g_{\mu\nu}}{n-1} = \frac{\delta R \sqrt{|g|}}{\delta g^{\mu\nu}} \qquad (3.7)$$

Where $\delta R \sqrt{|g|} / \delta g^{\mu\nu}$ represents the Lagrangian density of the gravitational field, which may improve the condition that the General Theory of Relativity does not assign a definite stress-energy tensor to the gravitational, leading to problems with local energy-momentum conservation, $R_{\mu\nu}$ is the Ricci tensor, $g^{\mu\nu}$ is the metric tensor, "n" is the number of spatiotemporal dimensions, and $|g|$ is the determinant of the metric tensor.

The variation with respect to the torsion tensor $T^{r\theta}{}_{\phi}$ yields the Cartan spin connection equations given by

$$T_{r\theta}{}^{\phi} + g_r{}^{\phi} T_{\lambda\rho}{}^{\rho} - g_{\lambda}{}^{\phi} T_{r\rho}{}^{\rho} = \frac{\delta R \sqrt{|g|}}{\delta T^{r\theta}{}_{\phi}} \qquad (3.8)$$

Consequently, the external spatiotemporal medium of the source of matter has no torsion, so, the external spatiotemporal geometry remains the same, due to the linearity, as typically described in the General Theory of Relativity. Thus, *torsion does not disperse, or travel as a wave*, even though a dispersion of torsion may be theoretically considered. *However, the torsion of the spatiotemporal medium may itself be a source of localized emergent time and gravity.*

It is possible to apply conservation laws for the Riemann-Cartan geometries since they have Lorentz symmetry as a local gauge symmetry. The metric and torsion tensors may act as independent variables providing the correct general law of conservation for the orbital and the intrinsic momenta, the total momentum, in the medium of the existing gravitational field.

A particle may travel in the trajectory of a curve, or a linear segment, through space-time, or radially at less than, or if massless, equal to, the speed of light "c". If an object such as a rotating black hole were to spin at nearly the speed of light, at nearly luminal intrinsic angular momentum, time on the event horizon of the black hole would dilate maximally to nearly tic-tac-zero. If the black hole were to travel simultaneously on a curved trajectory, it would travel at, or below, a relativistic speed due to its mass, through the spatiotemporal medium of our universe. Nonetheless, superluminal motion has been observed in the two opposing jets emanating from the core of a black hole, one jet moves toward the observer on Earth and another jet moves away from the observer.

Hence, with respect to our universe, time would dilate, and pass slower on an external frame of reference of the black hole due to curvature, or torsion, on its trajectory, or at an internal frame of reference of the spatiotemporal medium inside the event horizon of the black hole due to curvature or torsion, and intrinsic angular momentum, according to the ECFEs. *Consequently, time is nonlinear, curvature or torsion relate to linear, curvilinear, or radially inward or outward motion.*

Therefore, if the divine almighty creator of our universe stands outside the universe in the bulk where there is no divergence or convergence of the spatiotemporal medium, but the torsion, or intrinsic angular momentum, around the creator is not equal to zero, there is the possibility that the creator may be considered to be standing outside of our universal time, while the creator's interactive time may be adjusted as a just-in-perfect time. The bulk, in this context, represents an external volume that the creator can access, or create, for a divine purpose, and torsion may represent motion manifested by spiritual thought, since in complex physics, physical form may follow spiritual thought. (Nieves, 2020)

Chapter 3

The Quantum Complexity of Space-Time Inside the Event Horizon of a Black Hole.

§ 1. The emergence of quantum complexity.

Quantum States have an enormous capacity for spatiotemporal quantum complexity. It is possible to hypothesize that the spatiotemporal quantum complexity may be expressed by the sum of the amplitudes of spatiotemporal wavelets inside the event horizon of a black hole. This concept for a quantum field theory includes a sum over topologies. Amplitude may also refer to the geometric volume of a physical space, but it may also refer to the radius in a specific spatiotemporal direction. The spatiotemporal growth inside the event horizon of a black hole is hypothesized to be governed by complexity. Complexity is a physical paradigm to turn the inaccessible into physical reality.

$$|\Psi\rangle = \sum_{i=1}^{2^N} a^i |i\rangle \qquad (1.1)$$

Where a^i is a complex amplitude in 2^N quantum state coefficients of spatiotemporal wavelets, and Ψ is the spatiotemporal quantum complexity, or quantum mechanical superposition of states. The maximum entropy of a system of "n" spatiotemporal wavelets is "N", as the system comes to thermal equilibrium. There are two types of entropies to discuss black holes: The entropy of entanglement, or von Neumann entanglement entropy, a fine grained entropy, which is a measure of the degree of quantum entanglement between two subsystems constituting a two-part composite quantum system, remains constant under unitary time evolution, $t \equiv |\psi|^2 e^{\psi}$ in our physical reality. The other is the thermodynamic entropy, a coarse grained entropy, Boltzmann's entropy that arises from disorder, obeys the second law of thermodynamics, and it is larger than entanglement entropy. By the second law of thermodynamics for perturbations of dynamic black holes, the change of energy is related to a change in the acceleration of space-time, angular momentum, or electric charge. (Nieves, 2020)

The thermodynamic entropy *"N"* of the black hole corresponds to the area of the event horizon of a black hole. If the von Neumann entropy of an entangled system is zero that part of the system is isolated from the rest. The number of all possible interactions and arrangements of spatiotemporal wavelets or particles inside the event horizon of a black hole, or the storage capacity for the quantum information of a black hole, can be described from the outside as a quantum system with *"δ"* degrees of freedom at finite temperature in some specific space-times. There may be at least four types of black holes: stellar, intermediate, supermassive, and miniature. Each type would have a distinct area for its event horizon.

Consequently, in an evaporating entangled black hole, there are two ways of computing the entropy. Let us express the relationship between the entanglement entropy, or fine grained entropy, and the thermodynamic entropy, or coarse grained entropy, as follows:

$$S\left[\vec{\Psi}_a\right] \leq \sum_{i=1}^{2^N} S\left[\vec{\Psi}_{a^i}\right] \qquad (1.2)$$

Where $S\left[\vec{\Psi}_a\right]$ is the coarse grained entropy, and $S\left[\vec{\Psi}_{a^i}\right]$ is the fine grained entropy of entanglement in terms of 2^N quantum state coefficients of *"n"* spatiotemporal wavelets. The number *"λ"* of spatiotemporal waves with amplitude *"a"* of the coarse grained entropy consist of *"n"* spatiotemporal wavelets. It is reasonable to state that quantum gravitation identifies the microstates of fine grain entropy, while classical gravitation acknowledges the coarse grain entropy.

The arrow of time reverses inside the event horizon of a black hole to the arrow of ordentropy, as space-time converges in a contracting spatiotemporal medium. Matter turns into antimatter, space turns into time, time turns into space, and gravity turns into antigravity (gravitational repulsion), as volume converges, instead of diverging, toward the Planck scale. *The behavior of the fields of antimatter and antigravity, inside the event horizons of two nonconcentric entangled black holes, may initiate and contribute to the production of a black hole - white hole pair.*

$$m_p \to im_p \qquad (1.3)$$

$$t_p \to l_p \qquad (1.4)$$

$$l_p \to t_p \qquad (1.5)$$

Therefore, physical constants like c, G, $\hbar$, g, reverse their dimensions as follows:

$$G \to \frac{G}{ic^5} = \frac{t_p^3}{im_p \cdot l_p^2} \qquad (1.6)$$

$$\hbar \to \frac{i\hbar}{c^3} = \frac{im_p \cdot t_p^2}{l_p} \qquad (1.7)$$

$$c \to c^{-1} = \frac{1}{c} = \frac{t_p}{l_p} \qquad (1.8)$$

$$g_p \to \frac{t_p^3}{im_p \cdot l_p^2} \cdot \frac{im_p}{t_p^2} = \frac{t_p}{l_p^2} \qquad (1.9)$$

A spatiotemporal bridge, or its black hole, may grow or reduce its size. The volume or size of a spatiotemporal bridge may be described quantum mechanically in terms of the complexity of the quantum state of the system of spatiotemporal wavelets inside the event horizon of a black hole. *The maximal amount of complexity inside a black hole, during or after formation, cannot exceed the area of its outer event horizon.* The formation of a singularity may be described quantum mechanically in terms of the simplicity of the converging quantum entanglement entropy. The complementarity of complexity and simplicity emerges from the space-time duality. The interactions between complexity and simplicity are the essence and principles of the wave function of the spatiotemporal medium and the intricacy of its physical fields.

The spatiotemporal bridge cylindrical volume is the product of the temporal area A_t and a duration of proper time "τ_{PT}" in anti de sitter space-time inside the event horizon of the black hole. The complexity would be equal to the spatiotemporal volume, $V_t = A_t \cdot \tau_{PT[ADS]}$, of the spatiotemporal bridge divided by the product of the Newton gravitational constant and an anti de sitter length "l_{ADS}", a hypothetical static equation for complexity in five flat dimensions of an anti de sitter space-time, where dimensions in the numerator and in the denominator of the physical constants are multiplied by "i" and the speed of light raised to a power "c^n" to obtain an equivalent complex equation in the direction of the arrow of ordentropy, or in the direction of the reversed arrow of time.

$$C \simeq \frac{A \cdot \tau_{PT[ADS]}}{G \cdot l_{ADS}} \rightarrow \frac{A_t \cdot \tau_{PT[ADS]}}{[G/ic^5] \cdot t_{ADS}} \tag{1.10}$$

$$C \simeq \frac{i \cdot V_t \cdot c^5}{G \cdot t_{ADS}} \tag{1.11}$$

The hypothetical equation for complexity inside the event horizon of a black hole, in a (4 + 1) formalism for four-dimensional space-time, may be formulated as follows:

$$C \simeq \frac{i \cdot V_t \cdot c^5}{G} \tag{1.12}$$

Aside, a quantum length "l_q" may be expressed as a wavelength, $\lambda = hc/E = c/v$, where "v" is the frequency of a spatiotemporal wavelet. It is theorized that the dynamic equation for Planck quantum system complexity equals the product of the reduced Planck constant and $\sqrt{-1}$, divided by the speed of light cubed, times the number of states, e^{2^N}. The Planck acceleration "a_p" is approximately equal to $5.56072602 \times 10^{51}$ m/s^2. The Planck acceleration inside the event horizon of a black hole is given by the product of the Planck mass and $\sqrt{-1}$, times the number of states e^{2^N}, divided by the quantum Planck complexity. The value of Planck quantum complexity C_p is approximately $3.913938993 \times$

10^{-60} $(Kg \cdot s^2)/m$ times the product of $\sqrt{-1}$ and the number of states, e^{2^N}.

$$C_p \equiv \frac{i\hbar}{c^3} e^{2^N} = \frac{im_p \cdot t_p^2}{l_p} e^{2^N} = \frac{im_p}{a_p} e^{2^N} \tag{1.13}$$

Where $l_p = \sqrt{G\hbar/c^3}$ is the Planck length, $v_p = (G\hbar/c^3)^{\frac{3}{2}}$ is the Planck spatial volume, $t_p = \sqrt{G\hbar/c^5}$ is the Planck time, $v_p = (G\hbar/c^5)^{\frac{3}{2}}$ is the Planck temporal volume and $i = \sqrt{-1}$.

$$a_p = \frac{im_p}{C_p} e^{2^N} \tag{1.14}$$

1.1. The measurement of all things.

The mathematical tools, or the formulations of equations of physics, which are used to describe the fields of reality, and obtain their numeric values, illustrate that mass, matter, energy and space-time are all information. Therefore, *complexity is information*, and information is contrast. Even though, presently, the existence, or content, of that information, may be hidden, or inaccessible, to our perception or measurement. Information can neither be lost nor destroyed under the law of conservation of energy. The complexity is the storage media of the quantum information inside the event horizon of a black hole. A black hole serves, among other functions, as a quantum information repository, or memory, for the quantum information of knowledge about entanglement entropy, quantum states of entanglement, volume, mass, charge, kinematics, and temperature, of its physical existence.

1.2. A Planckian black hole.

"Things should not be more complicated than they have to be."

Hence, let us posit the formation of a Planckian black hole from a spatiotemporal spherical wave in the quantum foam of an entangled black hole right before saturation. In terms of the Hagedorn temperature, the entanglement entropy of "S_P" may be denoted as

$$T_H \triangleq \frac{1}{2\pi\sqrt{2}\ell_P} = \frac{1}{\sqrt{8\pi}\ell_P} \tag{1.15}$$

$$S_P \equiv -\frac{\hbar(\omega_P)}{T_H} \equiv -\sqrt{8\pi}\hbar(\omega_P)\ell_P \tag{1.16}$$

The entanglement entropy is shown with a negative sign to emphasize that the ordentropy occurs on the advanced wavelet of the converging spatiotemporal medium after saturation. The Hagedorn temperature "T_H" is the maximum temperature for a particular phase of matter, and going beyond that temperature requires a phase transition. In string theory, a Hagedorn temperature can be defined for strings when string energy exceeds the interaction forces between strings, disintegrating mass further into a plasma of strings.

1.3. The entanglement entropy toward a minimal area.

The entanglement entropy, or fine grained entropy, $S(\rho_s) \triangleq -Tr[\rho_s \ln \rho_s]$, has to be conserved in unitary quantum evolution. The natural log is used since saturation, inside the event horizon of an entangled black hole, is theorized to be a nonlinear natural process, while the "log" is a mathematical tool to linearize to some extent a nonlinear function. The natural log "ln" is the inverse function to the exponentiation of the natural growth "e" of divine creations. The string density matrix is "ρ_s," and "Tr" denotes the trace, where the entanglement entropy $S(\rho_s) = 0$ for pure states, and the dyad $\rho_S = |\rho_S\rangle\langle\rho_S|$, is in the form of a diagonal density matrix.

A diagonal density matrix with two non-zero terms or more on the diagonal constitute mixed states. A density matrix that has off-diagonal terms that are non-zero constitute superposition states; those states are referred to as coherent and the off-diagonal terms are referred to as coherences.

The entanglement entropy, or fine grained entropy, converges to a minimal area toward evaporation, but the Boltzmann Entropy, or coarse grained entropy, has maximally mixed states right before or at saturation. Boltzmann entropy is as if all information about what went into the black hole is coded on its outer event horizon. Nevertheless, there exists the possibility, or the need every now and then, that we decide to focus only on one part of the system and disregard the rest. The process is different when quantum properties have to be considered and will often correspond to trace over a specified subsystem.

As time emerges within an entangled black hole, through a Planck spatiotemporal radial length "l_P/c," the surface of the spherical wave may be conceptualized as a temporal diverging boundary "$4\pi t_P^2$." Right before the black hole reaches saturation, the spatiotemporal wave boundary expands encompassing an adjacent long string, then, the arrow of time reverses direction, and the spherical wave becomes a Planckian black hole as the highly excited entangled string collapses to its center as a converging point mass "M" with an approximate diameter of $\sqrt{M}$ and a temporal surface area of "$4\pi t_P^2$".

The infalling string is visualized as a single long string by a distant observer. The physical idea of the infalling string in General Relativity is observer-dependent. The surface turns into the boundary of a converging spatial surface and the interior of the spatiotemporal wave turns into a converging temporal volume of complexity. Consequently, let us consider a weakly coupled theory for the converging Planckian black hole. The spatiotemporal dimensions would be four dimensions with folded time according to a formalism of (3 + 1) from an unfolded formalism of (3 + 3), and four other theorized spatial dimensions that are compact, or curled up, unobservable and unmeasurable as of yet.

Since the size of the black hole is close to the string scale, $\ell_S \sim (M/m_p)\ell_p$, it is theorized that the string turns into a highly excited string. After formation, the entropy of the entangled string black hole may be denoted as

$$S_S \equiv -\sqrt{8\pi\hbar}(\omega_P)\ell_S \qquad (1.17)$$

For a weakly coupled theory, $g \ll 1$, $\ell_S = 1 = \sqrt{\alpha'}$, $\hbar(\omega_p)\ell_p/m_p^2 \simeq \ell_p^2 \sim g^2$, where $\sqrt{\alpha'}$ is the computed leading order corrections, to describe a highly excited string in thermodynamic equilibrium.

The thermodynamic entropy of a string black hole

$$\lim_{M \to \infty} M = dM \tag{1.18}$$

$$\frac{dV_{BH}}{dt} = \dot{a} = dA_{BH} \tag{1.19}$$

$$dA_{BH} = \frac{dM}{S_S} \tag{1.20}$$

As the string black hole mass converges and becomes more compacted, the volumetric acceleration increases, the complexity inside the event horizon increases, and the volume grows as well.

$$dM \equiv \frac{\hbar}{\frac{dA_{BH}}{dt}} \equiv \frac{\hbar}{\ddot{a}} \tag{1.21}$$

At formation, the mass dM, entropy S_S, and $\dot{a}$ of the entangled string black hole may be denoted as

$$dM = S_S \cdot \dot{a} = -\sqrt{8\pi\hbar(\omega_p)}\ell_S \cdot \dot{a} \tag{1.22}$$

The Schwarzschild Metric solved the Einstein Field Equations under the assumption of spherical symmetry for a black hole that is nonrotating and uncharged. The most obvious spherically symmetric problem is that of a point mass, or even a collapsed long string, with no orbital motion, only radial motion from or away from the center which is a radial distance "r_s" away.

The Schwarzschild radius of the event horizon of a black hole demarcates the boundary of the space-time duality that causes an object of mass to undergo irreversible spatiotemporal convergence after saturation.

As time turns into space and space turns into time during convergence, and $r_s > r$, $1-(r_s/r) < 0$, we have

$$c^2 d\tau^2 = -\left(1-\frac{r_s}{r}\right)c^2 dt^2 + \frac{dr^2}{c^2\left(1-\frac{r_s}{r}\right)} - \frac{r^2}{c^2}\left(d\theta^2 + \sin^2\theta d\varphi^2\right) \quad (1.23)$$

$$r_s = \frac{2\hbar(\omega_p)M\ell_p}{m_p^2 c^2} = \frac{2M\ell_p}{m_p} = 2\ell_s \quad (1.24)$$

Where "M" is a converging point mass, "m_p" is the Planck mass, "ℓ_s" is the string scale proper distance, "ℓ_p" is the initial radius of the throat and the Planck length, and "$d\tau^2$" is positive for time-like curves, in which case "τ" is the proper time.

$$c^2 d\tau^2 = -\left(1-\frac{2\ell_s}{r}\right)c^2 dt^2 + \frac{dr^2}{c^2\left(1-\frac{2\ell_s}{r}\right)} - \frac{r^2}{c^2}\left(d\theta^2 + \sin^2\theta d\varphi^2\right) \quad (1.25)$$

In the case of an entangled string black hole pair, a symmetrical Planckian spatiotemporal bridge which may be traversable to information and governed by quantum gravitation, may be expressed as

$$c^2 d\tau^2 = -\left(1-\frac{2\ell_s}{r}\right)c^2 dt^2 - \frac{d\ell_s^2}{c^2} - \frac{r^2}{c^2}\left(d\theta^2 + \sin^2\theta d\varphi^2\right) \quad (1.26)$$

$$c^2 d\tau^2 \approx -\frac{c^2 dt^2}{e^{\frac{2\ell_s}{r}}} - \frac{d\ell_s^2}{c^2} - \frac{r^2}{c^2}\left(d\theta^2 + \sin^2\theta d\varphi^2\right) \quad (1.27)$$

Where "r" is a function of the string scale proper distance "ℓ_s,"

$r \simeq \ell_S - M \ln(\ell_S/\ell_p)$, at a radial distance that is a long way from the throat between the two entangled string black holes in the above context. The function "r" is proportional to the proper string scale distance "ℓ_S."

The Planckian entangled string black hole travels on the advanced wavelet, with a tachyonic highly excited string, as the external spatiotemporal medium converges toward evaporation.

§ 2. *The monogamy of entanglement.*

The monogamy of quantum entanglement represents the fundamental property that the entanglement of numerous arbitrary quantum states cannot be freely shared.

For pure entangled states, A and B:

$$S(\hat{\rho}_{AB}) = 0 \qquad (2.1)$$

$$\hat{\rho}_{ABC} = \hat{\rho}_{AB} \otimes \hat{\rho}_C \qquad (2.2)$$

The pure entangled system is completely uncorrelated and statistically independent. Only two systems of pure states can be maximally entangled with each other, and it is impossible to add a third system that is entangled with the first two, or with one of the pure entangled states.

For a tripartite system, ABC, of entangled states:

$$|S(\hat{\rho}_{AB}) - S(\hat{\rho}_C)| \leq S(\hat{\rho}_{ABC}) \leq S(\hat{\rho}_{AB}) + S(\hat{\rho}_C) \qquad (2.3)$$

A composite system consisting of three constituents, a tripartite system, ABC, with the entropy of the composite state in the center, flanked by two inequalities.

The monogamy of entanglement has wide quantum mechanical applications from quantum cryptography to entangled black hole physics, where it plays a very important role in the security of quantum key distribution. Quantum key distribution is a secure communication method which applies a cryptographic protocol that includes quantum mechanical components. It can provide a shared random secret key between two parties that is only known to them, which can be used to encrypt or decrypt messages.

§ 3. The holographic principle of the event horizon of a black hole.

The holographic principle is a quantum mechanical idea that the event horizon of a black hole is kind of like a hologram, a two-dimensional piece of film or membrane, encoding quantum information of higher-dimensional character in a number of degrees of freedom on the boundary of the system, from the information stored inside the event horizon, or as information falls through the event horizon of the black hole. It is possible in principle to reconstruct three-dimensional reality from a two-dimensional hologram. Therefore, there may be a direct and instrumental correspondence between the quantum complexity information stored inside the event horizon of the black hole and the encoded hologram. As the black hole finally disperses, the hologram also disperses with it, and carries off the quantum information. (Lopresto, 2003)

The black hole radiant energy emittance is given by

$$\frac{dM_{BH}c^2}{dt} = \frac{\varepsilon \sigma T^4}{A_{BH}} = \frac{\sigma T^4}{4\pi R_S^2} \quad \text{(where } \varepsilon = 1 \text{ for a black hole)} \quad (3.1)$$

$$\frac{dM_{BH}}{dt} = \frac{\sigma T^4}{4\pi R_S^2 c^2} \quad (3.2)$$

$$\sigma = \frac{\pi^2 k_B^4}{60 \hbar^3 c^4} \quad (3.3)$$

$$T^4 = \left(\frac{\hbar c^3}{8\pi k_B G M_{BH}}\right)^4 = \frac{\hbar^4 c^{12}}{4096 \pi^4 k_B^4 G^4 M_{BH}^4} \quad (3.4)$$

$$R_S^2 = \left(\frac{2GM_{BH}}{c^2}\right)^2 = \frac{4G^2 M_{BH}^2}{c^4} \qquad (3.5)$$

$$\frac{dR_S}{dM_{BH}} = \frac{d\left(\frac{2GM_{BH}}{c^2}\right)}{dM_{BH}} = \frac{2G}{c^2} \qquad (3.6)$$

$$dV_{BH} = 4\pi R_S^2 \cdot \frac{dR_S}{dM_{BH}} = 4\pi \cdot \frac{4G^2 M_{BH}^2}{c^4} \cdot \frac{2G}{c^2} = \frac{32\pi G^3}{c^6} \qquad (3.7)$$

Substituting and solving for time in the black hole radiant energy emittance equation, we obtain

$$t_{DIS} = \left(\frac{8^4 \cdot 60}{4^2}\right) \frac{\pi G^2 M_{BH}^3}{\hbar c^4} = \frac{15360 \pi G^2 M_{BH}^3}{\hbar c^4} = \frac{480 c^2 M_{BH}^3}{\hbar G} \left(\frac{32\pi G^3}{c^6}\right) \qquad (3.8)$$

The time, in years, that the internal spatial volume and mass of a black hole take to disperse per solar mass $M_\odot$ is given by

$$t_{DIS} = \frac{480 c^2 V_{BH}}{\hbar G} \cdot \left(\frac{M_{BH}}{M_\odot}\right)^3 \simeq 2.1 \times 10^{67} \text{ years} \cdot \left(\frac{M_{BH}}{M_\odot}\right)^3 \qquad (3.9)$$

Where k_B is the Stefan-Boltzmann constant, R_S is the Schwarzschild radius, T is the Hawking absolute temperature in Kelvin, and a_P^2 is the Planck acceleration squared. The time that the internal temporal volume and mass of a black hole take to disperse is given by

$$\tau_{DIS} = \frac{480 c^2 \left(V_{BH}/c^3\right)}{\frac{i\hbar}{c^3} \cdot \frac{G}{ic^5}} \cdot \left(\frac{iM_{BH}}{iM_\odot}\right)^3 = \frac{480 c^7 V_{BH}}{\hbar G} \cdot \left(\frac{M_{BH}}{M_\odot}\right)^3 \qquad (3.10)$$

$$\tau_{DIS} = \frac{480 V_{BH}}{a_P^2} \cdot \left(\frac{M_{BH}}{M_\odot}\right)^3 = 5.082 \times 10^{109} \text{ years} \cdot \left(\frac{M_{BH}}{M_\odot}\right)^3 \qquad (3.11)$$

It is interesting to note, that when space turns into time, and time turns into space, inside the event horizon of a black hole after formation, it takes more temporal units of volume, a factor of c^5, or ~2.42×10^{42} m^2/s^2, to fill the same internal volume of units of space. Consequently, time converges very slowly compared to space. *Time expands radially to maximize the lifetime trajectory of particles through the existing geometry of space-time.*

Space-time may be classified in three kinds of geometries for the value of a cosmological constant "Λ".

Space-Time	Boundary	Λ
Asymptotically Anti de Sitter	Temporal ∞	< 0
Asymptotically de Sitter	Spatial ∞	> 0
Asymptotically Flat	Null	0

Table 1. Classifications of Spatiotemporal Geometries.

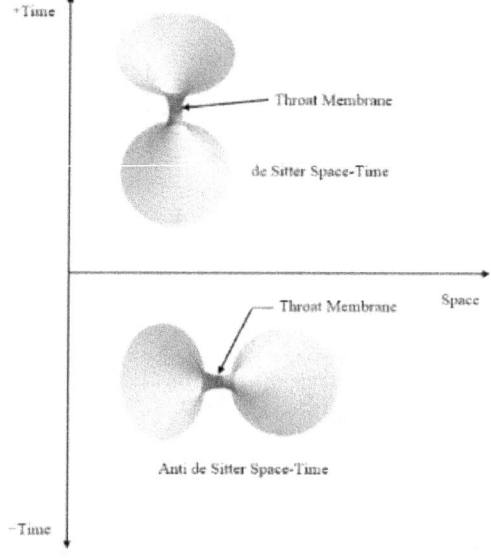

Figure 1. Spatiotemporal Bridges of "de Sitter" Space-Time and "Anti de Sitter" Space-time.

In the previous figures, the "de Sitter" space-time has a boundary in the past and another in the future while the "Anti de Sitter" space-time has a boundary on the side, as a radial direction goes to infinity.

The boundary of the AdS has a temporal boundary. Hence, the throat membrane of the AdS spatiotemporal bridge has a (2 + 1) formalism. The three-dimensional quantum field theories have the symmetry group SO(3,2) just like an AdS_4 space-time. Those theories with conformal symmetry and scaling are called Conformal Field Theory or "CFT_3". If the six-dimensional space-time were an asymptotical AdS space-time, then the throat membrane boundary would still have a (2 + 1) formalism.

It is interesting to note that Conformal Field Theories are invariant to spatial transformations such as translations, boosts, scalings, rotations, etc., that are conformal, that is, they preserve angles, regardless of scale. If there is scaling of the throat membrane, there would also be fixed points of flow or a renormalization group. A critical point is a special example of a phase transition. A strategy for physical processes involving many length scales is the renormalization group approach. The strategy involves tackling the physical process in steps, one step for each spatial scale. For critical physical processes, the technical solution is to find statistical averages over thermal fluctuations on all size scales. The renormalization group approach is to integrate out the fluctuations in sequence starting with fluctuations on a Planck scale and then proceeding to successively larger quantum scales until fluctuations on all scales have been averaged out. (Wilson, 1982)

§ 4. The Complex Ginzburg-Landau Equation (CGLE).

The Complex Ginzburg-Landau Equation was first derived in the studies of Poiseuille flow and reaction-diffusion systems. The equation is a physical theory used to describe superconductivity, which may also be given a general setting in the context of Riemannian geometry, which extends to quantum field theory and string theory. It is possible to hypothesize that the equations of motion and dispersion through the spatiotemporal medium at, onto, or near the throat membrane of an entangled black hole pair corresponds to the spatiotemporal quantum complexity expressed by the sum of the amplitudes of spatiotemporal wavelets inside the event horizon of an entangled black hole pair. Out the self-interactions of the spatiotemporal wavelets of quantum complexity, gravitation is the most important; it could be attractive from divergence and repulsive from convergence of the spatiotemporal medium of an entangled black hole pair.

Let us consider the case where $Im(\lambda) = 0$, and $\omega_c \neq 0$, so each mode of the spatiotemporal wavelets of matter fields and the complexity at or near the throat membrane, corresponds to traveling spatiotemporal wavelets with complex amplitudes equal to "A". If $Im(\lambda) = 0$, the unstable modes are growing in time for positive values of varying instability, but each mode is stationary in space.

Then, after rescaling the CGLE for the complex amplitude "A", we have the equations of motion and dispersion through the spatiotemporal medium onto and around the throat membrane of an entangled black hole pair.

The retarded dispersive wavelet method:

$$-\frac{\partial A}{\partial t} = (1 - i\alpha)\frac{\partial^2 A}{\partial r^2} + A - (1 - i\beta)|A^2|A \qquad (4.1)$$

The advanced dispersive wavelet method:

$$\frac{\partial A}{\partial t} = (1 + i\alpha)\frac{\partial^2 A}{\partial r^2} + A - (1 + i\beta)|A^2|A \qquad (4.2)$$

Where α and β are spatiotemporal parameters, k_c is the critical wave number, λ is the wavelength, and ω_c is the critical angular frequency. It is also interesting to note, that in the limit case of the above equation, α and $\beta \rightarrow \pm\infty$, reducing to the nonlinear Schrödinger equation, which possesses well known soliton, or solitary wavelet solution. A soliton, or solitary wave, is a regenerating wavelet packet that preserves its form while it travels at a constant velocity. Solitons are the result of wave interference of dispersive and nonlinear effects in the spatiotemporal medium. The solitary wavelet solutions are stable solutions of the nonlinear Schrödinger equation in which there is an ideal balance between dispersion and nonlinearity.

The solution for the dynamics of the retarded and advanced wavelets close to the threshold of the throat membrane may be denoted as,

$$u = u_0 + A(r,t)e^{ik_c r + i\omega_c t} + A^*(r,t)e^{-ik_c r + i\omega_c t} + \text{Higher Order Terms...} \quad (4.3)$$

Let us examine the two-dimensional CGLE using the Laplace operator "∇^2".

$$\frac{\partial A}{\partial t} = (1+i\alpha)\nabla^2 A + A - (1+i\beta)|A|^2 A \quad (4.4)$$

where A is a complex field. Apart from the two-dimensional spiral perturbation and phase turbulence, the above equation has a variety of coherent structures, like two-dimensional cellular structures known as quasi-frozen states. They appear in the form of quasi-frozen arrangements of spiral perturbations or vertices, and may provide a superconducting state on the throat membrane of an entangled black hole pair. It is possible to hypothesize that the event horizon of an entangled black hole pair may just be an impenetrable and symmetrical holographic membrane on which time is quasi-frozen, but information may still pass through it.

The symmetrical or central dispersive wavelet method:

$$\frac{\partial A}{\partial t} = \nabla^2 A + A - |A|^2 A \quad (4.5)$$

It is interesting to note that the equation above emerges naturally near any stationary supercritical subdivision if the system has translational invariance and is reflection symmetric $(x \to -x)$. Translational invariance implies that the equation above has to be invariant under the complex amplitude, $A \to Ae^{i\phi}$. When the instability is very small and supercritical, the nonlinearities saturate, or the complexity reaches saturation, so that the resulting patterns are above the threshold, have small amplitude and a wavelength close to $2\pi//k_c$.

Rewriting the symmetrical or central dispersive wavelet equation in a different form, we have

$$\frac{\partial A}{\partial t} = -\frac{\delta V}{\delta A^*} \quad (4.6)$$

$$V = \int_{r_1}^{r_2} \left(\left|\frac{\partial A}{\partial r}\right|^2 - |A|^2 + \frac{1}{2}|A|^4 \right) dr \qquad (4.7)$$

It is interesting to note that the volume "V" plays the role of a Lyapunov functional ($dV/dt < 0$). It is important to emphasize that the volume has a stationary phase winding solution given by

$$A = a_0 e^{iqr} \qquad (4.8)$$

$$q^2 = 1 - a_0^2 \qquad (4.9)$$

The solution describes steady state periodic patterns with the wave number slightly smaller ($q < 0$) or slightly bigger ($q > 0$) than the critical wave number k_c.

If the throat membrane emerges as a stationary supercritical surface for matter fields then the complexity of the wavelets would provide translational invariance, reflection symmetry, under the complex amplitudes of the wavelets in the spatiotemporal medium.

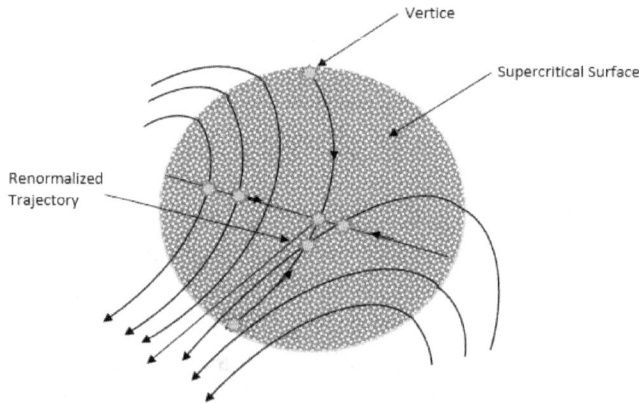

Figure 2. The Throat Membrane as a Supercritical Surface for Matter Fields.

The dynamics of matter fields of the SO(3,1) are equivalent to the dynamics of the matter fields of the CFT$_3$ according to the AdS/CFT correspondence, all other things being equal. It is theorized that the quantum gravity solutions of the "A Dynamic Theory of Space-Time", a six-dimensional quantum gravity theory with a (3 + 3) formalism, may simplify and resolve the existing quantum gravity difficulty at the throat membrane of an entangled black hole pair, by mapping the difficulty with folded time in a (3 + 1) formalism onto a CFT_3.

The Anti de Sitter Space/Conformal Field Theory (AdS/CFT) correspondence shows how the photons of Hawking radiation might be able to encode information about the interior of the black hole, thereby carrying that information back rapidly out into the universe at large after the black hole has dispersed halfway in a disproportionate amount of time depending on its size. The AdS/CFT duality states that an interacting system of particles within the event horizons of an entangled black hole pair corresponds to a higher dimensional gravitational theory. Any further quantum information falling into the black hole may be reflected, registered in its Hawking radiation, and emitted into the spatiotemporal medium. This conjecture may be the basis of a future holographic technology through quantum black holes for holographic teleportation of images, objects, and encrypted quantum information, leading to a cosmic internet for all entangled things, a cosmic entangled net. Any object that is already entangled may be accessible through this network. The coupled black holes may be able to send essential classical information that is totally scrambled, through a circuit of entanglement, teleporting the completely unscrambled quantum signal between black holes. Black holes are very fast quantum information scramblers across the surface of their event horizons. (Maldacena et alia, 2015)

The theoretical timescale, or scrambling time, t_σ, for non-linear time inside the event horizon of a black hole may be conjectured as:

$$t_\sigma = \frac{1}{2\pi T} \ln \tilde{M} \qquad (4.10)$$

Where $\tilde{M}$ is the actual ordentropy inside the event horizon of a black hole, T is the absolute temperature in Kelvin, and t_σ is conjectured as a theoretical bound on the growth rate of ordentropy in thermal quantum systems with a large number of degrees of freedom.

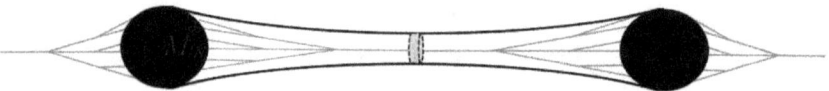

Figure 3. An Instance of Scrambling through an Entangled Black Hole.

$$\text{Divergent or Convergent Size} \sim e^{\mp\frac{2\pi}{\beta}t} \quad (4.11)$$

$$\text{Fully scrambled size} = N$$

After saturation, a particle would take "$t_{infalling}$" to travel from the event horizon or boundary to the stretched horizon a Planck distance away from the event horizon of the black hole.

$$t_{infalling} = \frac{\beta}{2\pi} \log N \quad (4.12)$$

The fastest scrambling time, or tachyonic scrambling, may be expressed as

$$t_{Tachyonic\ Scrambling} \geq \frac{\beta}{2\pi} \log S \quad (4.13)$$

The quantum scrambling behavior or process of a black hole takes local information in a deterministic way, $\lambda \leq 2\pi/\beta$, and spreads it across the entire event horizon of the black hole, generating quantum entanglement between distant regions. The decoding effect comes from the information signal or message going through the black hole's scrambling behavior or process, on the advanced wavelet, effectively unscrambling the quantum signal or message which arrives back at the point of entanglement of the black holes in a decoded form to be forwarded on the retarded wave to the other entangled black hole. Two entangled black holes may be conceived as an input/output register or buffer attached to an output display.

The event horizon may be conceptualized to be a scrambler with an encoder or a decoder, which are unitary transformations, so two event horizons and a throat horizon or membrane would function as three scramblers, each with an encoder-decoder, and a spherical display, in either direction of a black hole pair. A black hole is input-agnostic, or compatible with many types of codes, quantum transmissions, or transmitters. In principle, a quantum transmission with a single message should be enough to transmit an exact, and deterministic facsimile, but it would probably be unsuccessful. A high order quantum transmission provides a compromise that is exact, probabilistic, but requires multiple transmissions, to improve the probability of success with respect to an increase in the number of messages sent.

The input transmission, the signal states: $\Psi_1 \ldots \ldots \Psi_N$, may be sent to either black hole by a transmitter that is entangled to particles that are inside both black holes. The throat of the black hole pair provides a fixed communication channel medium between the black holes. The input transmission may be parallel between two black holes or sequential from a multi-entangled input case.

Input/Output Black Holes + Event Horizon Encoders/Decoders $\triangleq$ Holographic Displays

$$|M_{IN}\rangle + \frac{1}{\sqrt{2}}\left[\left(|E_1\rangle+|E_2\rangle\right)+\left(|BH_1\rangle+|BH_2\rangle\right)+\frac{\sqrt{2}}{\sqrt{3}}\left(|EH_1\rangle+|TM_2\rangle+|EH_3\rangle\right)+\left(|D_1\rangle+|D_2\rangle\right)\right]$$

$$\triangleq \left(1-e^{-C_p}\right)|-M_{OUT}\rangle \qquad (4.14)$$

Where E_1 and E_2 is the entangled encoders, D_1 and D_2 are entangled decoders, BH_1 and BH_2 are entangled black holes, EH_1, TM_2, EH_3, are black hole event horizons and a throat horizon or membrane between input and output, $\left(1-e^{-C_p}\right)$ is the probability of getting an exact facsimile at the receiving end, and M_{IN} and M_{OUT} are an input message and an output message.

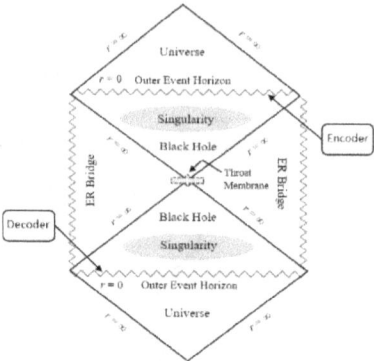

Figure 4. Quantum Entangled Black Holes.

The throat of the entangled black hole pair has both retarded and advanced wavelets on either side of the intermediate membrane for the quantum state signals of entanglement and causality.

Nonetheless, quantum entanglement breaks causality, and quantum entanglement brakes when the entangled particles decohere through their interaction with the environment.

The intermediate membrane encodes quantum information of three-dimensional character in a number of degrees of freedom on its boundary as information passes through the throat horizon of the black hole pair. If changes were made to a quantum message by the reversable procedural code of an encoder, then, the process of the encoder's procedural code may be reversed to recover the original quantum message.

$$\left| \Psi_{i_1} \right\rangle \otimes ... \otimes \left| \Psi_{i_N} \right\rangle \rightarrow Encoder \quad (4.15)$$

Entanglement may be regarded as a form of correlation in Quantum Mechanics. The correlation or dependence between two or more quantum states of massless quantum fields, if there were one or more quantum states of massless fields in each of the entangled black holes, with a throat between them that is quantum mechanically not eternal, that connects the two boundaries unobstructed by nondescript random harmonics, may be represented by

$$\operatorname{corr}(|BH_1\rangle, |BH_2\rangle)_{Throat} \triangleq \frac{1}{\sqrt{2}}\left(e^{\frac{(\ell_p - \ell_{Throat})}{\ell_p}} + ie^{\frac{(\ell_{Throat} - \ell_p)}{\ell_p}} \right) \quad (4.16)$$

Where $\pm(\ell_{Throat} - \ell_p)/\ell_p$ is the length of the throat minus the Planck length, over the Planck length, as the classical length of the throat would initially increase gradually up to saturation, or after matter fields enters the throat by $\sim 2\ln(\textit{throat radius})$, after any quantum mechanical recurrence, then decrease in the direction of ordentropy after saturation, as the correlation function would decrease until there was quantum mechanical recurrence, or the separation was completed, at $\ell_{Throat} \to \ell_p$. Furthermore, the minimal length of the throat of an extremal entangled black hole may be expressed as $\ell_{Throat} \triangleq N - (AdS_n \times S^n)$, where N is an index of extended supersymmetry, usually with an integer value of 2 or 4 supersymmetry of the supercoset model $AdS_n \times S^n$ of string theory, and "n" is the number of dimensions that in our context of the throat membrane, $n = 2$.

Integrability of a string background type model is a key feature that allows one to determine the string spectrum in non-trivial curved backgrounds, like the deformed AdS_n type of background and the S^n metric, of the supercoset model. The outside diameter of a spherical black hole may be defined as r_e, which equals the UV scale, in the framework of the UV self-complete quantum gravity of an extremal entangled black hole above the trans-Planckian quantum wavefoam.

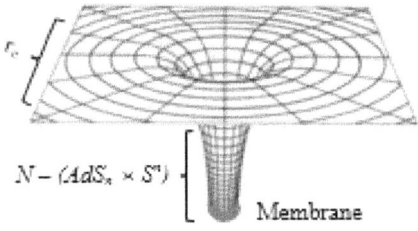

Figure 5. The Throat Membrane of an Extremal Entangled Black Hole.

Hence, if the throat thermodynamic process lasts enough time, the correlation of entangled quantum states of the black hole pair will return to what it was by less than, or equal to, a double exponential factor of the maximum quantum complexity/Planck quantum complexity, $\leq e^{e^{C_{MAX}/C_p}}$, the maximum thermodynamic entropy, Boltzmann constant, S_{ENT}/k_B, or the maximum ordentropy, Boltzmann constant, S_{ORD}/k_B, due to the Poincaré recurrence theorem.

The Poincaré recurrence theorem, a basic principle of quantum mechanics for a conservative system, states in the context of a certain dynamical black hole pair, a continuous state system, will, after a sufficiently long but finite time, return to a state arbitrarily close to its initial state. The length of time elapsed until the recurrence is the Poincaré recurrence time; the recurrence time may vary greatly depending on the exact initial continuous state, with the required degree of closeness. (Poincaré, 1890)

$$\frac{1}{\sqrt{2}}\left(e^{\frac{(\ell_p - \ell_{Throat})}{\ell_p}} + ie^{\frac{(\ell_{Throat} - \ell_p)}{\ell_p}}\right) \leq \frac{1}{\sqrt{2}}\left(e^{e^{\frac{C_{MAX}}{C_p}}} + ie^{ie^{\frac{C_{MAX}}{C_p}}}\right) \quad (4.17)$$

The intermediate membrane may be theorized as an instance of a left-to-right duality of the throat, or a conjectured relation between a higher-dimensional gravity theory, such as General Relativity, and a holographic two-dimensional quantum mechanical field theory. The membrane duality resolves the black hole information paradox to some extent, because it reveals the evolution of a black hole in a process consistent with quantum mechanics. As a metaphor, it is possible to suggest that the membrane duality is where General Relativity and Quantum Mechanics meet and shake chiral hands.

In terms of the thermodynamic entropy of one of the black holes in a black hole pair, we have

$$\frac{S_{BH}}{\pi} = \frac{A_{BH}}{\pi} \cdot \frac{c^3}{4G\hbar} = \frac{A_{BH}}{4\pi l_p^2} = \frac{A_{BH}}{A_p} \quad (4.18)$$

$$S_{BH} = \frac{A_{BH}}{A_p} \cdot \pi \qquad (4.19)$$

The area of the throat membrane of an entangled block hole pair is given by

$$A_p = \frac{A_{BH}}{S_{BH}} \cdot \pi \qquad (4.20)$$

$$S_{TM} = \frac{A_{TM}}{A_p} \cdot \pi = \frac{A_{TM}}{\frac{A_{BH}}{S_{BH}} \cdot \pi} \cdot \pi = \frac{A_{TM}}{A_{BH}} \cdot S_{BH} \qquad (4.21)$$

$$A_{TM} = \frac{S_{TM}}{S_{BH}} \cdot A_{BH} \qquad (4.22)$$

The area of the throat membrane is proportional to the product of the ratio of the entropies of the throat membrane to one of the entangled black holes in the pair and the area of the black hole. The thermodynamic entropy of the throat membrane is proportional to the product of the ratio of the areas of the throat membrane to one of the entangled black holes in the pair and the thermodynamic entropy of the black hole.

As the temperature of the throat is lower, the throat becomes larger, and there is a redshift factor. Gravitational redshift is the phenomenon where electromagnetic waves or photons travelling out of a gravitational well lose energy. This loss of energy corresponds to a decrease in the wave frequency and increase in the wavelength, known more generally as a redshift. As the spatiotemporal medium expands a certain volume, πe^2, a photon rides its spatiotemporal wavelet, tumbling as it moves, traveling in a helical trajectory, for a volume of $V_{ie} = \pi e^2 (r^2 + c^2 t^2)$.

The duration of the thermalization process of an entangled black hole occurs when the volume inside the event horizon reaches saturation. The event horizon's boundary is theorized to be thermodynamic, dissipative, with short-lived quantum excitations.

$$\tau_{THERM.} \triangleq \frac{\hbar}{S_{MAX\ ENT.} T_{BH}} \tag{4.23}$$

The quantum information, or complexity, sent to the receiving end of the entangled black hole pair is given by

$$e^{-C_p} = p_{BH} \tag{4.24}$$

$$C_p = -\ln p_{BH} \tag{4.25}$$

If the probability is very low or 0%, the quantum information received may be most of the message, or the entire message, an exact facsimile, but if the probability is very high or 100%, the quantum information received may be very limited or none.

The six-dimensional Einstein–Hilbert action represents an integral over time, taken along the path of the system between the initial time and the final time of the development of the system: where the integrand L_M is called the Lagrangian density term of kinetic energy (J/m^3) to describe any matter fields appearing in the six-dimensional theory. For the six-dimensional action integral to be well-defined, the trajectory has to be bounded in space and time with a (3 + 3) formalism.

As explained in "A Dynamic Theory of Space-time", we find that mass dilates as a result of the velocity of time decreasing as the object, or matter field, speeds up through space, and this effect increases the momentum of the object, or matter field, $dp / \partial H = -dt / \partial r$, and its kinetic energy, as the object, or matter field, travels against the increasing gravitational field about the mass opposite to the direction of motion. Consequently, the six-dimensional Einstein-Hilbert action integral is shown with a negative sign equivalent to a direction of 180°, or π, in the direction of the decreasing velocity of time. Time may be viewed as an intrinsic relative movement, and an emergent property, of the underlying dimensions of space (Nieves, 2020).

By the Corresponding Angles conjecture,

$$\text{Planck Quantum Complexity} \equiv -\frac{\text{Six} - \text{Dimensional Einstein} - \text{Hilbert Action}}{\pi \hbar} \quad (4.26)$$

As the temperature decreases, the free energy of the action becomes more independent of temperature. If the volume inside the event horizon of an entangled black hole is not already saturated, complexity is proportional to spatiotemporal action.

$$dC_p = \frac{-m_p c^2}{\pi \hbar} \cdot d\tau \quad (4.27)$$

$$\frac{dC_p}{d\tau} = \frac{-m_p c^2}{\pi \hbar} = \frac{-m_p \cdot l_p^2}{\pi \cdot \hbar \cdot t_p^2} = -\frac{\omega_p}{\pi} \quad (4.28)$$

$$-\frac{dC_p}{d\tau} = \frac{\omega_p}{\pi} \quad (4.29)$$

$$\pi \frac{dC_p}{d\tau} = \frac{\omega_p}{\pi} \quad (4.30)$$

Hence, "$dp \equiv (dP/c^2) \cdot dV \cdot (dC_p/dt)$" may express the momentum of the Planck quantum complexity that is proportional to the number of wavelet interference given by "$\Pi \equiv (dt/d\tau)(dC_p/dt) \equiv dC_p/d\tau$" which represents the complexity volume "dV" of a spatiotemporal maximal slice, within the event horizon of the black hole during formation, or within the throat of an entangled black hole pair during formation, where "t" is coordinate time and "τ" is proper time, and "dP" is the spatiotemporal pressure.

Aside, in the context of an entangled black hole during formation of its thermodynamic system, the Planck quantum complexity follows the arrow of time on the retarded wavelet, in the clockwise direction.

Hence, at 90° clockwise, the Planck frequency phasor would be at $-i$.

The Planck quantum complexity equals the angular Planck frequency "ω_P" of the holographic upper bound of the quantum complexity, of the spatiotemporal wavelets inside the event horizon of an entangled black hole during formation of its thermodynamic system, divided by π^2. (Rindler, 1969) The Rindler proper time "τ" equals to the product of the Rindler length and time coordinates, $(2\pi/i\beta)t$, or $-i\lambda t$, a Rindler spatiotemporal duration. Rindler proper time multiplied by temperature and entanglement entropy equals the Planck quantum complexity, $\tau \cdot T \cdot S = \tau \cdot E = C_p$, where T is absolute temperature, $i\beta$ is the imaginary part called the phase constant in *rads/meter*, representing the thermodynamic processing time, unless there are rare collisions, in hyperbolic Rindler coordinates, a hyperbolically accelerated reference frame.

$$\frac{dC_p}{d\tau} = \frac{\omega_P}{\pi^2} \tag{4.31}$$

$$\frac{dC_p}{d\tau} = \frac{2f_p}{\pi} = \frac{f_p}{90°} = -if_p \tag{4.32}$$

$$\therefore \frac{dC_p}{d\tau} \equiv \frac{df_p}{d\theta} \tag{4.33}$$

$$\frac{d\theta}{d\tau} \equiv \frac{df_p}{dC_p} \tag{4.34}$$

$$\frac{dC_p}{df_p} = \frac{1}{\frac{d\theta}{d\tau}} = \frac{1}{v_\tau} = \frac{1}{\sqrt{1-\frac{v^2}{c^2}}} = \gamma = \frac{\mu_0}{\omega_s} \tag{4.35}$$

$$\frac{v^2}{c^2} = \frac{g_p l_p}{c^2} = 1 - \frac{1}{\left(\frac{dC_p}{df_p}\right)^2} \tag{4.36}$$

The velocity of Planck quantum complexity per linear Planck

frequency "f_p" equals the Lorentz factor of special relativity, or the ratio of permeability of free space-time to the spatial angular frequency of the spatiotemporal wavelets during the formation of an entangled black hole. Hence, it is reasonable to theorize that there may be Special and General Relativity factors at work with Planck quantum complexity, inside the event horizon of an entangled black hole during its formation. (Nieves, 2020) The hypothesized spatiotemporal quantum complexity expressed by the sum of the amplitudes of spatiotemporal wavelets inside the event horizon of a black hole, may also be defined as the sum of the quantum perturbations over interacting wavelet vertices. The work in the Planck energy scale, at which the quantum perturbations, or the gravitational effects of "n" interfering spatiotemporal wavelets are relevant, may be denoted as follows:

$$W_p = n(\hbar\omega_p) = n\sqrt{\frac{\hbar c^5}{G}} \simeq n \times 1.9561 \times 10^9 \, J \qquad (4.37)$$

Let us posit about the phenomenon of the Richarson-Dushman Law, which was first investigated by O. W. Richardson in 1900 and 1901. The following variation of the Richardson-Dushman equation for emission in a vacuum in the absence of external electric fields, is given by the charge density in current per unit of surface area, $\rho_q = AT^2(1-\bar{r})e^{B/k_B T}$, where A is the Richarson constant, $A = 4\pi q k_B^2 \sigma (m/\sqrt{2})/h^3$, σ is a correction factor for surface matter field, where $\sigma = 1$ for an ideal conductive holographic membrane, $\bar{r}$ is the coefficient of reflection from the potential barrier, or membrane, at the boundary of the emitter for thermoelectrons with different energies, B is the zero-bias mean barrier height, and the absolute temperature (T) of the emitting material. (Wolf, 1995, and Smithells, 1997)

The thermionic emission in the throat of an entangled black hole pair, or the Edison effect, is the flow of charged particles called thermions from a charged membrane surface, caused by the thermal vibrational energy overcoming the electrostatic forces holding electrons to the membrane surface matter field in a spatiotemporal medium.

In the context of an entangled black hole pair, for the linear frequency "f_p" of the charge density, ρ_q, as Planck charge/(Planck time × Planck length²), of a thermionic emission from the surface of the holographic throat membrane, or a holographic event horizon, to the energy of the thermodynamic entropy "S" as a work function up to saturation, we have

$$\rho_p = \frac{q_p}{t_p \cdot \ell_p^2} = \frac{2^{3/2} \pi \cdot q_p \cdot \sigma}{T^2 \cdot \ell_p^2} \cdot T^2 \cdot (1 - \bar{r}) \cdot e^{-\frac{S}{k_B}} \qquad (4.38)$$

$$\frac{1}{s} = 2^{3/2} \pi \sigma (1 - \bar{r}) e^{-\frac{S}{k_B}} \qquad (4.39)$$

If the coefficient of reflection is zero and $\sigma = 1$,

$$\frac{1}{s} = 2^{3/2} \pi e^{-\frac{S}{k_B}} \qquad (4.40)$$

$$df_p = 2^{3/2} \pi e^{-\frac{S}{k_B}} = 2^{3/2} \pi \frac{d(cycles)}{dt_p} = 2^{3/2} \pi \frac{dT_{Periods}}{dt_p} = 2^{3/2} \pi \frac{d\theta}{dt_p} \qquad (4.41)$$

$$\frac{d\theta}{dt_p} = \frac{1}{e^{\frac{S}{k_B}}} = v_\tau = \sqrt{1 - \frac{v_r^2}{c^2}} \qquad (4.42)$$

It is interesting to note that for fermionic emission, Planck quantum complexity is directly proportional to the exponential of the thermodynamic entropy ratio "S/k_B", This conjecture is a direct correspondence between Quantum Mechanics, General Relativity, and Information.

$$QM \equiv GR \equiv \dot{I} \qquad (4.43)$$

$$e^{\frac{S}{k_B}} \equiv \frac{1}{\sqrt{1-\frac{v_r^2}{c^2}}} \equiv \frac{dC_p}{df_p} \qquad (4.44)$$

Therefore, it is possible to state that Quantum Mechanics, General Relativity, and Information, are so closely related that it would be difficult to disentangle them. However, the holographic conjecture of an entangled black hole pair allows a researcher to realize that QM, GR, and I, are theories of exactly the same marvel of physical reality; they are all parts of the quantum, small, and large scale of the structure of the universe, which developed thermodynamically, especially from a lower complexity state to a higher complexity form. The duality and evolution of an entangled black hole exemplify the universal development and growth. Hence, Duality and Gravitation are emergent at large N theories, in higher dimensional space-time. Some quantum systems that interact strongly may have physical gravitational duals.

In a thermionic emission, the holographic throat membrane would supply heat from some electrons with at least the minimal energy required to overcome the attractive force holding them in the structure of the conductive medium of the membrane. This minimal energy, called the work function, would be a characteristic of the emitting holographic membrane and the state of the distribution of charges on its surface.

§ 5. Gravitation inside the event horizon of an entangled black hole pair.

For an entangled black hole pair, a gravitational mode may become strongly coupled at low energies, like for charged and supersymmetric extremal black holes, with attributes like $M \geq Q$, or even $M \sim Q$. For strongly coupled low energies, we have

$$\text{Effective Interaction Strength} \triangleq \frac{\text{Radius of Curvature}}{(\sqrt{2})\ell_p} \qquad (5.1)$$

Entropies of large black holes match beautifully with index

computations. There is more to a black hole than its entropy, for example, correlation functions, and the computation of AdS_D correlators in two dimensions, $D = 2$. The throat and its holographic membrane may be regarded as a boundary gravity mode that behaves quantum mechanically and merits to be treated in exact terms.

The Planck constant "h" is the defining constant of quantum mechanics, as well as "G", the Newton gravitational constant, is the defining constant of gravitation. Hence, any physical theory which combines Special Relativity, Quantum Mechanics, and Gravitation will presumably involve Planck units, Special Relativity and Quantum Mechanics lead to a very tight structure. Planck units simplify some recurring algebraic expressions in the laws of physics, and are distinctly relevant in quantum gravitational theory. For example, during the inflationary period the quantum gravitational effects are of order $\sim 10^{-5}$. The cosmological constant would be different at the Planck scale, $\Lambda_p = 1/\ell_p^2$. (Nieves, 2020) So, Einstein's cosmological constant may be expressed as

$$\Lambda_{abcd} = |C_{abcd}|/\ell_p^2 \text{ where } C_{abcd} = -\kappa \Lambda_{abcd} \text{ and } \kappa = 8\pi G/c^4. \quad (5.2)$$

The Planck quantum gravity "g_p" may be expressed in terms of the Planck acceleration and the square of the ratio of the Planck linear frequency to the Planck quantum complexity in the Lorentz factor. Hence, the Planck acceleration and the square of the ratio of the Planck linear frequency to the Planck quantum complexity are directly proportional to the emergence of an effective Planck quantum gravity within the event horizon of an entangled black hole during formation and spatiotemporal divergence.

$$g_p = \frac{c^2}{l_p} - \frac{c^2}{l_p \left(\frac{dC_p}{df_p}\right)^2} = \frac{l_p}{t_p^2} - \frac{l_p}{t_p^2}\left(\frac{df_p}{dC_p}\right)^2 = \left(\sqrt{\frac{c^7}{\hbar G}}\right) - \left(\sqrt{\frac{c^7}{\hbar G}}\right)\left(\frac{df_p}{dC_p}\right)^2 \quad (5.3)$$

$$g_p^+ = a_p - a_p\left(\frac{df_p}{dC_p}\right)^2 \qquad 1 \geq \frac{\partial f_p}{\partial C_p} \geq 0 \quad (5.4)$$

Where $a_p = 5.5608 \times 10^{51}$ m/s^2 is the Planck acceleration during formation. After formation and during spatiotemporal convergence, the space and time units would be interchanged, the unit of linear Planck frequency would be spatial *(1/m)*, and the units of Planck acceleration, or Planck quantum gravity, would be spatiotemporal *(−s/m²)*. Thus, the Planck quantum gravity may be expressed as,

$$\overline{g_p} = -\left(a_p - a_p \left(\frac{\partial f_p}{\partial C_p} \right)^2 \right) = a_p \left(\frac{\partial f_p}{\partial C_p} \right)^2 - a_p \qquad 1 \geq \frac{\partial f_p}{\partial C_p} \geq 0 \qquad (5.5)$$

Consequently, when the Planck acceleration "a_p" or the linear Planck frequency "δf_p" equals zero, the Planck quantum gravity goes to zero and the entangled black hole disperses. At saturation, $g_p \sim \pm a_p$, the Planck quantum gravity transposes from spatial acceleration to temporal. At *"N"* finite saturation corrections, the internal Planck quantum gravity in a Planck volume of an entangled black hole may be denoted by

$$g_p \equiv \left(\frac{1}{N} \right) \frac{\partial^2 V_p^{1/3}}{\partial t_p^2} \equiv \left(\frac{1}{N} \right) \frac{8\pi \hbar (\omega_p)}{m_p \ell_p} \equiv \left(\frac{1}{N} \right) \frac{8\pi c^2}{\ell_p} \equiv \text{Planck Quantum Gravity} \qquad (5.6)$$

$$\text{Number of Finite Saturation Corrections} \equiv \frac{1}{\text{Number of Quantum Gravity Corrections}} \qquad (5.7)$$

The Planck quantum complexity is always greater than zero after formation, even when the entangled black hole disperses, because the spatiotemporal wavelets may still converge or diverge in the spatiotemporal medium. The form of the spatiotemporal medium is very malleable and may be deformed by curvature, torsion, energy density or pressure, where the measurable deformation depends on the amount of energy density or pressure.

The holographic upper bound on the quantum complexity is expressed in terms of the conserved quantities of the holographic boundary at the event horizon of an entangled black hole. The boundary theory of an entangled black hole is a theory of strongly interacting particles. The state of the boundary of an entangled black hole follow the laws of physics present at the boundary. It is theorized that the external black hole gravitation is in the strong field range, ~ c^2/r_{BH}, if the entangled black hole is theorized to rotate near or at the speed of light "c", and the entangled particles would not travel through time, only through outside space. The numerous particles on the boundary of the black hole correspond to the boundary thermodynamic entropy and temperature proportionally to the spatial motion of the particles nearly or at tic tac zero due to redshift, from the point of view of an outside observer. (Nieves, 2020)

The ordentropy of the black hole, S_{ORD}, is theorized to be equal to the sum of the average products of the probability of each message at the receiving end with the quantum information or complexity contained in each message. (Shannon, 1948)

$$S_{ORD} = (1-p_1)C_{P_1} + (1-p_2)C_{P_2} + ... + (1-p_n)C_{P_n} \qquad (5.8)$$

$$S_{ORD} = -(1-p_1)\ln p_1 - (1-p_2)\ln p_2 + ... - (1-p_n)\ln p_n \qquad (5.9)$$

The encoded classical message could pass through an inverter before entering the input black hole encoder, to get an exact facsimile that would first pass through the intermediate membrane or horizon of the throat, come out into the second black hole as an inverted facsimile, to be projected on the second event horizon as an exact facsimile. The teleported information may be retrieved at the far end of the black hole pair as a faster than light communication. As technology improves, an input tomography device scans an object to create an exact set of cross-sectional facsimiles, a technique for displaying an image of a cross section through an object using X-rays or ultrasound to be teleported, to reconstruct a three-dimensional duplicate of the internal structures and the external features of the object, while preserving its isometry. At the receiving end, a three dimensional printer may recreate the object. The input tomography scan would involve greater complexity within a black hole pair. The tomography scan operation allows for the teleportation of an object through a pair of entangled black holes with existing technology, separated by a galactic or universal distance.

A quantum bit of information that would fall into one of two entangled black holes would be registered in the other. So, an entanglement spatiotemporal bridge may be used to manipulate black hole quantum information. Hence, it may be possible to use quantum information and teleportation to make a spatiotemporal bridge that is traversable. (Gao, Jafferis and Wall, 2019)

The geometry of an entanglement spatiotemporal bridge may cause gravitation or antigravitation to emerge from positive/negative spatiotemporal pressure, positive/negative curvature, and torsion from the spins of particles. Entanglement, Geometry and Gravitational Energy are strongly coupled. Energy or mass causes the spatiotemporal medium to curve and to emerge from the wave function, while entanglement leads to General Relativity. Entanglement leads to Geometry, Geometry is Emergent, Geometry leads to Positive/Negative Curvature and Torsion, and Positive/Negative Curvature and Torsion lead to Gravitation or Antigravitation. Therefore, Entanglement leads to Gravitation or Antigravitation. The patterns of entanglement are intently related to the spatiotemporal structure. Moreover, entanglement describes geometry and establishes spatiotemporal energy, to specify that positive/negative pressure equals negative/positive energy density, on each direction of the arrow of time. However, gravitation or antigravitation may emerge from more than a single source.

If the entanglement between two black hole regions is reduced, the entangled black holes may start pulling apart showing that entanglement is necessary for the formation of the throat for a black hole - white hole pair. Hence, if the entanglement were reduced to zero between the two black holes then the black hole and white hole would separate, and the throat connecting would dissipate. On the other hand, the entropy of entanglement contains a contribution from the spatiotemporal expansion of the medium. Therefore, since the spatiotemporal wavelet is complex in its nature, it may be a particle, a wave, or a particle-wave, depending on its quantum states. As the space-time inside the event horizon of a black hole contracts, the spatiotemporal pressure increases versus the outside spatiotemporal pressure, expanding the diameter of the black hole, and the black hole grows. The density ρ_{a^i} of the amplitudes of spatiotemporal wavelets inside the event horizon of a black hole, or inside its spatiotemporal bridge, is directly proportional to its spatiotemporal pressure, P_{BH}.

The convergence process may be the mechanism of string compactification or spaghettification of matter, accreted by a black hole, the tidal effect caused by strong gravitational fields inside the event horizon of a black hole as matter free-falls towards the singularity of a black hole. Nothing inanimate ends, or becomes nothingness, if it collides with the singularity of a black hole, not even time or space. The external black hole volume exists toward its future, the internal volume evaporates toward its past, only to return evaporated into its external future. After formation, the event horizon of a black hole is a special spatiotemporal turning point. The winding mode of a self-gravitating string in thermodynamic equilibrium in the throat of an entangled black hole pair may produce light or photon pair production. As matter free-falls in the black hole, it heats up, and emits high energy photons or other thermal radiation. Could these high energy photons initiate the antigravitational process of the conjugate white hole?

$$P_{BH} = \rho_{a^i} \qquad (5.10)$$

$$P_{BH} = \frac{im_p/c^2}{V_{BH}} \qquad (5.11)$$

Where V_{BH} is the temporal volume inside the event horizon of the black hole, and im_p is the Plack mass of a singularity.

What does the growing volume inside the event horizon of a black hole correspond to, at the Planck level?

It is theorized that the maximal volume inside the event horizon of a black hole may grow exponentially as a function of the number "N" of quantum state coefficients of "n" spatiotemporal wavelets inside the event horizon of a black hole at maximum entanglement entropy and directly proportional to the Planck quantum complexity. Thus, as the black hole forms, the volume inside the event horizon of a black hole grows due to its increasing complexity, to its maximum entanglement entropy. Moreover, the roles of the spatial and temporal coordinates are exchanged, so it is very contrasting to define a concept of the inside volume that is consistent with what we consider as volume on the outside of the event horizon of a black hole. General Relativity describes the interior volume of a black hole as warped space-time. Hence, any conventional integral over that internal region of the volume could be ill-defined. Even though the singularity is characterized as having infinite density, inside the event horizon of a black hole, it becomes a future temporal instant, not a spatial location. *The duration of the divergence or convergence of the spatiotemporal medium may be conceptualized as spatial or temporal depending on the direction of the arrow of time.*

The outside size of a black hole is related to the radius of its event horizon but its inside volume is related to the length contraction and the infinite curvature within its event horizon. Inside a black hole, there are radial space-like curves of infinite length, while radial time-like curves have proper time or clock time.

How does the quantum complexity of spatiotemporal wavelet interference give rise to diverging or converging spatiotemporal volume inside the event horizon of a black hole?

The maximum quantum complexity provides a measure of the number of interactions that would be required to recuperate the initial quantum state of a black hole at the initial moment that it formed. After the black hole's formation, spatiotemporal wavelets inside the event horizon of a black hole start to decouple with each other, very gradually becoming less scrambled, and converge. As the arrow of time reverses, the complexity of a black hole continuously and gradually decreases due to spatial wavelength contraction, spatiotemporal wavelets dissociate, or disentangle, as time turns into space, and space turns into time, and ordentropy increases. Black holes may accrete matter and space-time over time, which may assist the decreasing entropy of a black hole depending on how much entropy has been added to its internal volume. It is not that the Boltzmann's entropy of a black hole always increases from whatever configuration the thermodynamic system is at, but the next configuration of the system is most likely to have an increased entropy. It is also valid to state that if the system is at any configuration point among random configurations, the previous configuration was most likely to have an increased entropy. Hence, it may be more correct to state that entropy almost always increases when it follows the retarded arrow of time, or that entropy almost always decreases when it follows the advanced arrow of time.

The energy per absolute temperature inside the event horizon of the black hole can be theoretically estimated, depending on the direction of the arrow of time. The following thermodynamic entropy-and-ordentropy dyad denotes the same relation that characterizes the retarded wavelet and the advanced wavelet of the spatiotemporal medium, as will be discussed later.

$$(|S_{BH}\rangle\langle -S_{ORD}|)_{BH} = \frac{1}{\sqrt{2}}\left(\frac{e}{T_{BH}^2}\right)^2 (|S_{ENT}\rangle + \langle i^2 S_{ORD}|)$$

$$= \frac{1}{\sqrt{2}}\left(\frac{e}{T_{BH}^2}|S_{ENT}\rangle + \frac{e}{T_{BH}^2}[|i^2 S_{ORD}\rangle]^\dagger\right) \quad (5.12)$$

Consequently, the spatiotemporal divergence-or-convergence mechanism inside the event horizon of a black hole may play an important role in the creation of spatiotemporal bridges between entangled particles that may have fallen into two or more distinct black holes that would eventually connect through the throat. The entangled particles would follow the arrow of ordentropy. The arrows of entropy and ordentropy are conjugates in the same way that the retarded arrow and the advanced arrow are conjugates during the divergence or convergence of the spatiotemporal medium. The black hole progresses through divergence and convergence inside its event horizon until it finally disperses.

The maximal spatial volume, $V_{BH_{max}}$, of a black hole, as it forms, may be expressed as

$$V_{BH_{max}} = v_p e^{2^N} = \left(\frac{G\hbar}{c^3}\right)^{\frac{3}{2}} e^{2^N} \quad (5.13)$$

Where 2^N is the minimal complexity of the quantum state coefficients of "n" spatiotemporal wavelets inside the event horizon of a black hole at "maximum thermodynamic entropy", and v_p is the Planck volume, 4.2217×10^{-105} m^3. The maximum thermodynamic entropy of a system of "n" spatiotemporal wavelets is "N", where the number of states is e^{2^N} entirely.

The minimal temporal volume, $V_{BH_{min}}$, of a black hole, after its formation, may be expressed as

$$V_{BH_{min}} = v_p e^{-2^M} = \left(\frac{G\hbar}{c^5}\right)^{\frac{3}{2}} e^{-2^M} \qquad (5.14)$$

The equation, $N + M = 100\% = 1$, is the sum of the thermodynamic entropy and the ordentropy inside the event horizon of a black hole; so, the minimal volume may also be expressed in terms of the thermodynamic entropy, $M = 1 - N$, as follows

$$V_{BH_{min}} = v_p e^{-2^{(1-N)}} = \left(\frac{G\hbar}{c^5}\right)^{\frac{3}{2}} e^{-2^{(1-N)}} \qquad (5.15)$$

Where 2^M is the maximum complexity of the quantum state coefficients of m spatiotemporal wavelets inside the event horizon of a black hole at "maximum ordentropy". The maximum ordentropy of a system of "m" spatiotemporal wavelets is M, where the number of states is e^{-2^M}, or $e^{-2^{(1-N)}}$, entirely.

It is interesting to note that the third law of thermodynamics for black holes states that a temperature of absolute zero Kelvin of an object is at its minimal possible thermodynamic entropy, or its maximum ordentropy. As a result of this law, it is thought that an object cannot be cooled to absolute zero Kelvin. A black hole may be freezing cold inside the event horizon, at one millionth of a degree above absolute zero Kelvin, if its ordentropy is maximal. Outside the event horizon, the temperature of a black hole is astonishingly hot. The temperature increases as the outside volume of the black hole decreases. Furthermore, absolute temperature and time are reciprocal during ordentropy.

$$T_{BH} \sim \frac{1}{t_{BH}} \qquad (5.16)$$

The Big Bang theory postulates that in the beginning the universe came into being from a single, infinite density and gravity, "hot" singularity, with an absolute temperature ~ 10^9 K, and space and time did not exist. However, it is known that "hotness" involves the passage of time in the direction of the retarded wavelet. In the beginning, there were photons, while electrons and positrons annihilate to produce more photons, which are particles of light. It is interesting to ask the following rhetorical questions, could scientists have postulated the incorrect Bing Bang theory? Could it be possible that the Big Bang singularity was infinitely "cold" and time and/or space did not exist yet in or around the singularity? Is it possible that the Black Hole thermodynamic process and the Big Bang singularity thermodynamic process are chiral, or a mirror image of each other? If that scenario were possible, the singularity might be treated classically instead of quantum mechanically in the trans Planckian medium under the Grand Unified Law of Physics, or the Law of All There Is or All That Will Be. Did the universal singularity come into being inside an infinite All There Was as a holographic reality of complexity? Is it all about Physical Laws or Grace? (NASB, 1995)

Therefore, as the saturation of a black hole decreases during ordentropy, the absolute temperature of the black hole decreases toward zero as time turns into space and space turns into time. The volume of the black hole may decrease slowly toward evaporation. As absolute temperature decreases, the passage of the duration is on the advanced wavelet, not on the retarded wavelet of the saturation process. Toward the end of the evaporation process the temperature of the black hole is astonishingly cold.

As the absolute temperature of the black hole decreases toward zero, time dilates and space contracts, as the retarded wavelet and the advanced wavelet slow down toward standstill at an arbitrary point in the spatiotemporal medium, as a quantum phases transition. As the name quantum phases transition suggests, molecules in their state of matter may be much closer together than in the solid, liquid, plasma or gaseous state, undergoing a reduction in volume, area, or length, and a dilation in time, because the spatiotemporal distances contract to yield a more compact molecular structure or atomic scale. The nuclei may be treated classically while the electrons may be treated quantum mechanically, or a simulation in a classical computer may lead to quantum entanglement breakthroughs in quantum mechanical group renormalization for the throat membrane of an entangled black hole pair. *The "temperature-temporal correlation" is a phenomenon that involves scaling symmetry of the complexity wavelet spatiotemporal medium based on converging or diverging radial distance.* Symmetries provide more constrained spatiotemporal geometrical structures and stronger coupling.

The retarded quantum complexity of the black hole during formation is given by

$$C_{RET} = \frac{S_{ENT} \cdot T_{RET} \cdot \Delta t_{RET}}{c^3} e^{2^N} = \frac{\hbar}{c^3} e^{2^N} \qquad (5.17)$$

Where S_{ENT} is the thermodynamic entropy during formation, T_{RET} is the retarded temperature, Δt_{RET} is a period of retarded proper time.

The advanced quantum complexity of the black hole after it formed is given by

$$C_{ADV} = \frac{S_{ORD} \cdot T_{ADV} \cdot \Delta t_{ADV}}{c^3} e^{-2^M} = \frac{i\hbar}{c^3} e^{-2^M} \qquad (5.18)$$

Where S_{ORD} is the ordentropy after formation of the black hole, T_{ADV} is the advanced temperature, and Δt_{ADV} is a period of advanced time.

Retarded quantum complexity and the number of states of thermodynamic entropy inside the event horizon of a black hole are proportional, and advanced quantum complexity and the number of states of ordentropy are reciprocal. Consequently, black holes eventually dissipate.

The encrypting of the spatiotemporal medium inside the event horizon of a black hole is created from the extremely subtle correlations of the spatiotemporal wavelet interference that define quantum complexity. Complexity builds up geometry, or deconstructs geometry, as long as time passes, depending on the direction of the arrow of time. Complexity may be described as a dynamic evolution governed by the emergence of time and a principle of action describing how a physical system has changed over time. The internal volume of a black hole is an instance of dynamic spatiotemporal geometry. The geometrical features inside the event horizon of an entangled black hole may be rendered by the existence of an entanglement pattern in the theoretical state of the boundary. The spatiotemporal medium emerges independently from the particles living on the boundary or their entanglement. The spatiotemporal medium plays an essential role in determining the structure of entanglement. Space-time is emergent and fundamental. What is the source of emergence or physical animation, the physical laws of grace or the divine laws of physics? Grace reminds a researcher consistently that the love of physics may be a virtual part of the love of grace.

Chapter 4

The Laws of Complexity.

§ 1. The first law of quantum complexity for entangled black holes.

The first law of quantum complexity for entangled black holes calculates the difference in complexity between a spatiotemporal wavelet interference at a fixed point, as a first reference state $|\psi_{FRS}\rangle$, and a set of spatiotemporal wavelet interferences at a subsequent point, that is a second point state $|\psi_{SPS}\rangle$, when the second point state is a small-scale perturbation $|\psi_{SPS} + \delta\psi\rangle$ of the first reference state, and where t_P parametrizes the quantum state, and the Planck time or the position, inside the event horizon, as an independent quantity, along the path. (Dowling and Nielsen, 2008)

$$\delta C = C|\psi_{SPS} + \delta\psi\rangle - C|\psi_{SPS}\rangle \qquad (1.1)$$

$$\delta C^{(1)} = p_w \cdot \delta\tau^w \big|_{tp=1} - p_w \cdot \delta\tau^w \big|_{tp=0} \qquad (1.2)$$

If the temporal position of the first reference state is taken as fixed, the second term is negligible, and we obtain,

$$\delta C^{(1)} = p_w \cdot \delta\tau^w \big|_{tp=1} \qquad (1.3)$$

$$p_w \equiv \frac{\delta(im_p v_P^{-2})}{\delta v_P^{-1}} \qquad (1.4)$$

The first law equation is the inner product of the momentum p_w in the direction of the first reference state wavelet and the Planck time $\delta\tau^w$, inside the event horizon, that may or may not be orthogonal and vanishing, to the direction that the original wavelet is following. The variation of the first order coefficient C' reduces to the boundary term and to the leading order. The first law equation of quantum complexity for entangled black holes is a balanced equation, for the change in the quantum complexity, that any spatiotemporal wavelet notion of complexity should satisfy.

When $\delta C^{(1)} = 0$, the second order coefficient variation C'' is evaluated to determine the change in the quantum complexity as follows:

$$\delta C^{(2)} = \frac{1}{2} \delta p_w \cdot \delta \tau^w \big|_{tp=1} \qquad (1.5)$$

Therefore, the first law equation of complexity may be recapped as

$$\delta C = p_w \cdot \delta \tau^w \big|_{tp=1} + \frac{1}{2} \delta p_w \cdot \delta \tau^w \big|_{tp=1} \qquad (1.6)$$

It is interesting to note that the change, or variation δC, of the complexity is completely determined by the quantum state of the spatiotemporal wavelet interference at the terminal end point. These variations are between nearby temporal geodesics. Hence, the first law describes the variation of the quantum complexity for entangled black holes when the second point state is perturbed and the result only depends on quantum state information at the endpoint of the trajectory of the spatiotemporal wavelets. The first law of quantum complexity for entangled black holes may be examined with a holographic framework for complexity which can describe coherent states of the spatiotemporal wavelets with small amplitudes.

§ 2. The second law of quantum complexity for entangled black holes.

The second law of quantum complexity for entangled black holes states that if the volume inside the event horizon of a black hole is not already saturated, the quantum complexity of the volume will increase with an overwhelming probability toward its maximum value. Hence, gravitation may also emerge as a force of acceleration from the geometry of complexity, as well as an entropic force corresponding to the second law of thermodynamics.

If $\delta C = 0^+$ and the thermodynamic entropy S_{BH} is maximal,

$$p_w \cdot \delta\tau^w |_{tp=1} = -\frac{1}{2}\delta p_w \cdot \delta\tau^w |_{tp=1} \qquad (2.1)$$

$$p_w = -\frac{1}{2}\delta p_w \qquad (2.2)$$

$$-ip_w = \frac{i\delta p_w}{2} \qquad (2.3)$$

The quantum complexity of the volume and the thermodynamic entropy inside the event horizon are correlated. The growth of quantum complexity toward saturation confirms a statistically enormous probable occurrence inside the event horizon of an entangled black hole. The second law of quantum complexity follows the arrow of entropy. Once the thermodynamic entropy is maximal, the volume is saturated, and the momenta of the spatiotemporal wavelets and the variation of quantum complexity become temporal in nature. Space becomes time, and time becomes space.

§ 3. The third law of quantum complexity for entangled black holes.

The third law of quantum complexity for entangled black holes states that if the volume inside the event horizon of a black hole is saturated after reaching thermal equilibrium, the quantum complexity of the volume will decrease with an overwhelming probability toward its minimum value.

If $\delta C < 0$ and the ordentropy S_{ORD} is increasing,

$$\delta C = -p_w \cdot \delta\tau^w |_{tp=1} - \frac{1}{2}\delta p_w \cdot \delta\tau^w |_{tp=1} \qquad (3.1)$$

$$-\delta C = ip_w \cdot \delta\tau^w |_{tp=1} + i\frac{1}{2}\delta p_w \cdot \delta\tau^w |_{tp=1} \qquad (3.2)$$

The decreasing quantum complexity of the volume and the ordentropy inside the event horizon are correlated. The reduction of quantum complexity toward deficiency confirms a statistically enormous probable occurrence inside the event horizon of an entangled black hole. The decreasing variation in quantum complexity is temporal as space-time converges. Consequently, the temporal distance $\delta\tau^w$ is a Planck proper time inside the event horizon of the saturated black hole.

§ 4. The fourth law of quantum complexity dispersion for entangled black holes.

The fourth law of quantum complexity for entangled black holes states that if the complexity of the volume inside the event horizon of an entangled black hole has been depleted, the event horizon of the black hole will be dissipated, and its matter, energy, and spatiotemporal complexity will be dispersed to the surrounding spatiotemporal medium.

If $\delta C = 0^-$, $C_p = 0^-$, and $V_{BH} \simeq 0^-$, the ordentropy S_{ORD} inside the event horizon reached its maximum; subsequently, the complexity and the volume of the entangled black hole are zero.

$$C_p = -\ln p_{BH} = 0^- \quad (4.1)$$

$$S_{ORD} = -(1-p_1)\ln p_1 - (1-p_2)\ln p_2 + ... - (1-p_n)\ln p_n = 0^- \quad (4.2)$$

The hypothetical principles of quantum complexity for black holes, white holes, and spatiotemporal bridges are as follows:

1. Quantum complexity may be stable, unique, and minimal where there is spatiotemporal pressure equilibrium and thermal equilibrium between the spatiotemporal medium inside and outside a black hole, a white hole, or a spatiotemporal bridge, but there is no expansion or contraction of the black hole, white hole, or spatiotemporal bridge. This principle follows the condition that the thermodynamic entropy, the ordentropy, the temperature, and the spatiotemporal pressure are at equilibrium for a black hole, a white hole, and a spatiotemporal bridge.

2. Quantum complexity increases for black holes, or

spatiotemporal bridges in an expanding spatiotemporal medium outside the event horizons or throats, or a contracting spatiotemporal medium inside the event horizons or throats. White holes are theorized to have expanding spatiotemporal medium inside the event ektropí. This principle follows the arrow of entropy outside the event horizon of a black hole, and the arrow of ordentropy inside the event horizon of a black hole or a spatiotemporal bridge in a "black hole - white hole" pair, or the arrow of entropy inside the event ektropí of a white hole. This principle implies that the total energy of a single black hole is conserved, not created or destroyed, but may be converted from one form of energy to another.

3. Quantum complexity decreases for black holes, or spatiotemporal bridges in a contracting spatiotemporal medium outside the event horizons or throats, or an expanding spatiotemporal medium inside the event horizons or throats. This principle follows the arrow of ordentropy outside the event horizon of a black hole and the arrow of entropy inside the event horizon of a black hole or a spatiotemporal bridge in a "black hole - white hole" pair, or the arrow of ordentropy inside the event ektropí of a white hole. The total energy of a single white hole is conserved, not created or destroyed, but may be converted from one form of energy to another.

4. The arrow of quantum complexity follows the arrow of time, and the principles of quantum complexity follow the laws of thermodynamics. The principles of quantum complexity are time reversal invariant. Quantum complexity asymmetry in time is a salient and consequential feature of the thermodynamic qualities of the spatiotemporal medium.

§ 5. The retarded and advanced wavelets of the arrow of time at a black hole.

Let us envision a film editor running a famous digitized moving picture on three screens simultaneously. The left screen shows the film going forward in time (the retarded screen), the right screen shows the film going backward (the advanced screen), and the center screen is a superposition of both images at any instant of time. The editor can stop the film at any time, or at any scene, where the image at the center screen (the present image) will always be very sharp without any fuzziness at all. On the present image, the protagonist is sitting at the bar in the lobby of a hotel having a glass of sparkling water to quench his thirst, when he turns to look at his lovely wife walking toward him, smiles, puts his glass down on the counter, to speak with her. As the editor restarts the left and right images, on the right screen, the protagonist seems to be lifting the glass from the counter, the center screen stays the same, and on the left screen, the protagonist stops holding the glass on the counter, to speak to his lovely wife. This film story is a metaphor for how the interaction of the retarded matter wavelet and the advanced matter wavelet are superimposed along a world volume, to create a spatiotemporal present that is time reversal invariant, depending on the direction of the arrow of time, in the production of a black hole - white hole pair, in one of the solutions to the Einstein Field Equations of General Relativity.

Let us posit a thought experiment where an observer, not necessarily a person or a lifeform, crossing the event horizon of a black hole in free-fall notices no distinctive attribute, while the arrow of time reverses to the arrow of ordentropy, as space-time converges in a contracting spatiotemporal medium. In such a scenario, the matter, energy, and space-time, reverse toward its past, creating a mirror image or a backward moving picture, as material contracts toward the Planck scale and space-time converges. As the observer falls through the event horizon, the observer's matter and energy leaves an imprint, or scan, thermalized and recorded, on the stretched event horizon surface, going through a Lorentz transformation, that produces an exact multidimensional duplicate of the observer, in a reversed free-fall, inside the spatiotemporal medium of the event horizon. The observer's imprint may be reflected as lasting Hawking radiation to the external spatiotemporal medium, which may preserve the information of the free-falling observer throughout the surrounding and slowly diverging space-time. It is interesting to mention the black hole firewall paradox, or AMPS paradox, that old Hawking radiation inside the black hole may be entangled with new Hawking radiation while the very late Hawking radiation gets entangled with the late incoming Hawking radiation partners due to the fact that entanglement may be monogamous, which may be contradictory. The firewall resolution requires a violation of Einstein's equivalence principle, which states that a free-falling object of mass is indistinguishable from an object of mass floating in empty space-time.

§ 6. The principle of equivalence of the event horizon of a black hole.

The principle of equivalence of an event horizon of a black hole states that if an object of mass or a particle were on a spatiotemporal manifold, or inside a membrane, that converges as much as the outside spatiotemporal medium diverges, the object of mass or particle would appear to be symmetrical, in its spatiotemporal bubble at tic tac zero, with an effective rest mass, from the point of view of a stationary observer in the external spatiotemporal medium. The equation of motion of a particle on the throat membrane would be subject to the addition of a retarded motion of divergence and an advanced motion of convergence that results in a symmetrical wavelet effect.

The event horizon is an imaginary boundary that demarcates the external spatiotemporal medium of a black hole from the internal spatiotemporal medium. Either spatiotemporal medium may be diverging or converging according to the laws of physics present at either side of the demarcation boundary. After formation of the black hole, the retarded arrow of time and the advanced arrow of time are duals at the boundary. If an object of mass, a particle, or pure energy, not being disintegrated by a boundary firewall, or held on the temporal boundary surface, passes through the event horizon of the black hole, will exist in the interior spatiotemporal medium, and may remain entangled to other objects of mass, particles, or pure energy, in its universe. At the boundary of the event horizon after formation, there may be a relativistic gravitational field in a diverging spatiotemporal medium, on the retarded side, and a tachyonic gravitational field in a converging spatiotemporal medium, on the advanced side. Any object of mass, particle, or pure energy will be confined within the spatiotemporal medium or as part of the singularity of the black hole until its evaporation, unless the black hole is part of an entangled black hole pair, which may provide an exit through a throat bridge to its conjugate white hole and its external medium.

§ 7. A quantum continuous equation for the decoherence and the evaporation of an entangled black hole pair.

The quantum of time, or chronon, provides a possibility to formulate the phenomenon of quantum decoherence due to the evaporation of a charged particle at the spatiotemporal throat membrane, such as the excited state of a bound electron to facilitate the utilization of a finite difference equation instead of a differential equation, in three distinct temporal methods: a retarded wavelet, a symmetrical wavelet, and an advanced wavelet, in a Lorentz invariant theory. The retarded wavelet method, applicable at the throat membrane, or at an event horizon, of an entangled black hole pair, provides a fundamental quantum mechanical process to exchange with, or evaporate a quanta of energy, $(2\pi n)\hbar v$, to the surrounding spatiotemporal medium.

The chronon is denoted by

$$\theta_0 \triangleq \frac{q^2}{6\pi\varepsilon_0 m_0 c^3} \qquad (7.1)$$

Since the value of the chronon depends on the associated particle's internal properties of charge "q" and mass "m", the nature of the particle being considered must be specified. The numerical value of a chronon is approximately of the same order of magnitude as the time it takes light to travel across the classical diameter of an electron, ~ 10^{-24} sec. (Caldirola, 1953, 1980)

The temporal quantum theory resolves the following conditions for exact relativistic solutions: free electron motion, an electron under an electromagnetic pulse, a hyperbolic motion, and for nonrelativistic approximate solutions: an electron in a constant and uniform magnetic field, an electron moving along a straight line under the action of an elastic restoring force, and an electron under the action of time-dependent forces.

Hence, space-time is an approximate but effective conceptualization from particles existing on the spatiotemporal manifold or membrane of a black hole. The following finite difference equations are equations of motion for electrons in three distinct methods.

There are three four-dimensional distinct methods of Schrödinger equations to perform a quantification:

The retarded wavelet method:

$$i\frac{\hbar}{\tau_p}\left[\Psi\left(\vec{r},\vec{t}_r\right)-\Psi\left(\vec{r},\vec{t}_r-\tau_r\right)\right]=\hat{H}\Psi\left(\vec{r},\vec{t}_r\right) \qquad (7.2)$$

The symmetrical wavelet method:

$$i\frac{\hbar}{2\tau_p}\left[\Psi\left(\vec{r},\vec{t}_r+\tau_r\right)-\Psi\left(\vec{r},\vec{t}_r-\tau_r\right)\right]=\hat{H}\Psi\left(\vec{r},\vec{t}_r\right) \qquad (7.3)$$

The advanced wavelet method:

$$i\frac{\hbar}{\tau_p}\left[\Psi\left(\vec{r},\vec{t}_r+\tau_r\right)-\Psi\left(\vec{r},\vec{t}_r\right)\right]=\hat{H}\Psi\left(\vec{r},\vec{t}_r\right) \qquad (7.4)$$

All these spatiotemporal methods transform into a continuous equation when the fundamental interval of proper time "τ_r" goes to zero. Time is regarded as a continuous variable, but the system evolution along its world line may be regarded as incremental on the Planck scale. "r" is a resultant spatial distance, $r=\sqrt{x^2+y^2+z^2}$, "t_r" is a resultant temporal duration, and $t_r=\sqrt{t_x^2+t_y^2+t_z^2}$, and "$\tau_p$" is a resultant Plank proper time. For every spatial dimension, there is a conjugate temporal dimension. (Nieves, 2020)

Aside, it is interesting to note that the forward temporal direction, in a three-dimensional cartesian coordinate system, is the negative direction of time in the lower third quadrant away from the origin, for time is emergent.

The quantum of time in the finite difference equation decrements in that direction toward the future, the imaginary direction, or the direction of incoming future space, or the quantum of time increments, toward the positive temporal direction, toward the past, in the first upper quadrant away from the origin, in the existing space.

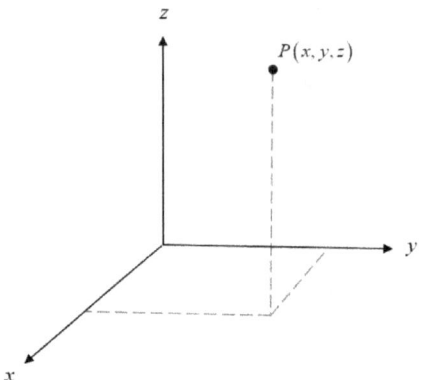

Figure 1. A Three-Dimensional Cartesian Coordinate System.

The following equations demonstrate the six-dimensionality of the Hamiltonian arguments with a (3 + 3) formalism:

$$i\frac{\hbar c}{\tau_p}\Psi(\vec{r},\vec{t}_r) \equiv i\frac{\hbar c}{\tau_p}\left[\Psi(\vec{r}_x,\vec{r}_y,\vec{r}_z,\vec{t}_x,\vec{t}_y,\vec{t}_z)\right] \qquad (7.5)$$

$$\hat{H}c\Psi(\vec{r},\vec{t}_r) \equiv \hat{H}c\Psi(\vec{r}_x,\vec{r}_y,\vec{r}_z,\vec{t}_x,\vec{t}_y,\vec{t}_z) \qquad (7.6)$$

The six-dimensional Hamiltonian equations of motion may be denoted as:

The retarded wavelet method:

$$i\frac{\hbar c}{\tau_p}\left[\Psi(\vec{r},\vec{t}_r) - \Psi(\vec{r},\vec{t}_r - \tau_r)\right] = \hat{H}c\Psi(\vec{r},\vec{t}_r) \qquad (7.7)$$

The symmetrical wavelet method:

$$i\frac{\hbar c}{2\tau_p}\left[\Psi\left(\vec{r},\vec{t}_r+\tau_r\right)-\Psi\left(\vec{r},\vec{t}_r-\tau_r\right)\right]=\hat{H}c\Psi\left(\vec{r},\vec{t}_r\right) \quad (7.8)$$

The advanced wavelet method:

$$i\frac{\hbar c}{\tau_p}\left[\Psi\left(\vec{r},\vec{t}_r+\tau_r\right)-\Psi\left(\vec{r},\vec{t}_r\right)\right]=\hat{H}c\Psi\left(\vec{r},\vec{t}_r\right) \quad (7.9)$$

These six-dimensional equations of motion for an electron can be easily simplified to four-dimensional equations of motion for an electron, with a (3 + 1) formalism when time is folded in a resultant temporal duration, but that mathematical operation does not make time one-dimensional, or linear. Time is three-dimensional, or nonlinear, like its conjugate space.

The value of the electron velocity is supposed to jump from only certain discrete positions, or certain probabilistic orbitals, along its world line; these discrete positions being such that the electron takes a proper time "τ_r" for traveling from one position to the next.

The symmetrical wavelet method is obtained by adding together the retarded and the advanced wavelet methods, made up of exactly similar wavelets facing each other showing symmetry. The solutions of the retarded and advanced wavelet methods exemplify an entirely distinct behavior. The evaporative behavior of an electron shows up in the nature of the retarded method with an exponential decrease. The electron radiates with intrinsically evaporative solutions in the case of the retarded method. The advanced method shows the opposite behavior since it follows a reversed arrow of time and an exponential increase in condensation, for an electron which can absorbs energy from the external spatiotemporal medium, particles and fields.

The fine distinction between continuous and discrete time, or proper time, is very delicate and precise, but bona fide, as it depends on one's conceptualization. The solutions for a four-dimensional Hamiltonian, explicitly independent of continuous time, based on discrete Planck and discrete proper time, have a general form given by $f(\vec{r})$ with a diverging, converging, or symmetrical in terms of each of a set of independent functions, the eigenfunctions of $\hat{H}$, which are the solutions to the given finite difference equation.

$$\Psi(\vec{r},\vec{t}_r) = \left[-e^{i\pi} + i\frac{\tau_p}{\hbar}\hat{H}\right]^{-\frac{t_r}{\tau_r}} f(\vec{r}) \qquad (7.10)$$

$$\Psi(\vec{r},\vec{t}_r) \equiv \frac{1}{-e^{-i\pi} \cdot f(\vec{r})^{-1}} + \frac{1}{\left(i\frac{\tau_p}{\hbar}\hat{H}\right)^{\frac{t_r}{\tau_r}} \cdot f(\vec{r})^{-1}} \qquad (7.11)$$

Thus, it is clear that the damping factor depends critically on the proper temporal value "τ_r" of Cardirola's chronon. The greater the discrete proper time, the lesser the damping factor. (Nieves, 2020)

For these Hamiltonians, $dp/\partial H = \mp dt/\partial r$, the effect of the quantification shows up primarily in the frequencies related to the time-dependent term of the wavelet function.

The six-dimensional Hamiltonian equations for divergence, convergence, or symmetrical acceleration, $f''(\vec{r})$, during the motion of an electron may be denoted as:

The 2nd Order of a Spatiotemporal Field

$\equiv$ *The Coupling Energy Strength* $\times$ *The Field Source*

The second order divergence acceleration method of the retarded wavelet:

$$i\frac{\hbar c}{\tau_p}\left[\frac{\Psi(\vec{r},\vec{t}_r+2\tau_r) - 2\Psi(\vec{r},\vec{t}_r+\tau_r) + \Psi(\vec{r},\vec{t}_r)}{\tau_r^2}\right] = \frac{\hat{H}c\Psi(\vec{r},\vec{t}_r)}{\tau_r^2} \qquad (7.12)$$

$$i\frac{\hbar}{\tau_p}\left[\Psi(\vec{r},\vec{t}_r+2\tau_r) - 2\Psi(\vec{r},\vec{t}_r+\tau_r) + \Psi(\vec{r},\vec{t}_r)\right] = \hat{H}\Psi(\vec{r},\vec{t}_r) \qquad (7.13)$$

The second order convergence acceleration method of the advanced wavelet:

$$i\frac{\hbar c}{\tau_p}\left[\frac{\Psi(\vec{r},\vec{t}_r)-2\Psi(\vec{r},\vec{t}_r-\tau_r)+\Psi(\vec{r},\vec{t}_r-2\tau_r)}{\tau_r^2}\right]=\frac{\hat{H}c\Psi(\vec{r},\vec{t}_r)}{\tau_r^2} \quad (7.14)$$

$$i\frac{\hbar}{\tau_p}\left[\Psi(\vec{r},\vec{t}_r)-2\Psi(\vec{r},\vec{t}_r-\tau_r)+\Psi(\vec{r},\vec{t}_r-2\tau_r)\right]=\hat{H}\Psi(\vec{r},\vec{t}_r) \quad (7.15)$$

The second order central acceleration method of the symmetrical wavelet:

$$i\frac{\hbar c}{\tau_p}\left[\frac{\Psi(\vec{r},\vec{t}_r+\tau_r)-2\Psi(\vec{r},\vec{t}_r)+\Psi(\vec{r},\vec{t}_r-\tau_r)}{\tau_r^2}\right]=\frac{2\hat{H}c\Psi(\vec{r},\vec{t}_r)}{\tau_r^2} \quad (7.16)$$

$$i\frac{\hbar}{2\tau_p}\left[\Psi(\vec{r},\vec{t}_r+\tau_r)-2\Psi(\vec{r},\vec{t}_r)+\Psi(\vec{r},\vec{t}_r-\tau_r)\right]=\hat{H}\Psi(\vec{r},\vec{t}_r) \quad (7.17)$$

§ 8. About the quantum information of a free-falling object of mass, energy and space-time into a black hole.

As the quantum information of the matter, energy, and space-time, of the observer in free-fall through the two-dimensional membrane of the event horizon of a black hole, the quantum information of three-dimensional character is encoded in a number of degrees of freedom on the boundary of the system. The entanglement attributes of the fundamental degrees of freedom are closely connected to the spatiotemporal medium. It is theorized that the imprinting mechanism, which thermalizes and records the quantum information on the stretched surface of the event horizon of a black hole, is the Lorentz transformation of the retarded wavelet and the advanced wavelet by the reversal of the arrow of time. The observer's dual imprint on the advanced wavelet enters the internal spatiotemporal medium of the black hole forward while the observer's genuine imprint on the retarded wavelet exits backward to the external spatiotemporal medium above the external surface of the event horizon of the black hole. The lasting external image of the observer exits backwards as time diverges very slowly from the point of view of an external observer. The lasting external image is layered with genuine quantum information of three-dimensional character which emit Hawking radiation to the external spatiotemporal medium. The Hawking radiation concept is similar to the production of primordial fluctuations during the inflationary period. Therefore, the particle-wave duality of the observer serves twofold purposes, preserving genuine quantum information that is available to the universe and providing internal dual quantum information to the internal spatiotemporal medium or matter inside the black hole.

Inside the event horizon of a black hole, it may be theorized that space turns into time, and time turns into space. Per "A Dynamic Theory of Space-Time," time emerges from space, and then more space emerges from time, in an expanding spatiotemporal medium outside the event horizon of a blackhole. In the low energy limit with slowly varying gravitational fields of larger waves in the region of smaller curvature, "A Dynamic Theory of Space-Time" reduces to the Theory of General Relativity. Hence, space-time may converge in a contracting spatiotemporal medium inside the event horizon of a black hole, or diverge in an expanding spatiotemporal medium outside the event horizon of a black hole. When the arrow of time reverses direction along each temporal dimension within the event horizon of a black hole, entropy decreases providing greater order. Temporal wavelets exhibit universal time-reversal symmetry. Ordentropy follows the reversal of the arrow of time, the higher the ordentropy the higher the uniformity or regularity inside the event horizon of a black hole. The arrow of ordentropy is a consequence of the reversal of the arrow of time in the space-time of gravitationally bound systems or objects with negative specific heat capacity. (Nieves, 2020)

Conversely, inside the event horizon of a black hole where space-time contracts, pressure increases, the conjugate roles of space and time reverse, space remits into time, and then more time remits from space.

As pressure increases inside the event horizon of a black hole, time dilates, increasing the amplitude of the spatiotemporal wavelets, contracting the spatial wavelength, and increasing the frequency.

Consequently, the Planck quantum complexity C_p increases correspondingly.

The static volumetric Planck acceleration (m^3/s^2) inside the event horizon of a black hole during formation may be expressed as

$$\frac{1}{\ddot{v}_p} = \frac{m_p}{\hbar c} \tag{8.1}$$

$$\ddot{v}_p = \frac{\hbar c}{m_p} \tag{8.2}$$

The static volumetric Planck acceleration during formation is approximately $1.452617641 \times 10^{-18}\ m^3/s^2$.

The dynamic volumetric Planck deceleration ($-m^3/s^2$) inside the event horizon of a black hole after formation is given by

$$\ddot{v}_p = -\frac{i\hbar c}{im_p c^3} e^{-2M} = -\frac{\hbar}{m_p c^2} e^{-2M} \tag{8.3}$$

Let us consider a concentric "black hole - white hole" pair formation at a spatiotemporal locality. The spatiotemporal region of the black hole would be represented by its retarded state $|\Psi_r\rangle$ and the spatiotemporal region of the white hole by its advanced state $\langle\Psi_a|$ which become maximally entangled through coherence. As the retarded state and the advanced state move out of phase, a maximally entangled "black hole - white hole" pair emerges, the black hole travels on the retarded wavelet toward the future and the white hole travels on the advanced wavelet toward the past. If the black hole and the white hole are measured on the same basis, the same outcome is obtained. Hence, the black hole would exist toward the future and the white hole would exist toward the past of the same universe as implied in one of the solutions to the Einstein Field Equations of General Relativity.

$$\left(|\Psi_r\rangle\langle\Psi_a|\right)_{BH\cdot WH} = \frac{1}{\sqrt{2}}\left(|\Psi_r\rangle_{BH\cdot WH} + \langle\Psi_a|_{BH\cdot WH}\right) = \frac{1}{\sqrt{2}}\left(|\Psi_r\rangle_{BH\cdot WH} + \left[|\Psi_a\rangle\right]^\dagger_{BH\cdot WH}\right) \quad (8.4)$$

Aside, the dyad, or the linear operator for the outer product, $\left(|\Psi_r\rangle\langle\Psi_a|\right)_{BH\cdot WH}$, is a single row matrix, or row vector, multiplying a single column matrix, or column vector, giving a single numerical value in a finite-dimensional complex vector space, to find expectation values of momentum or location.

Aside, let us review and propose the quantum mechanical "cloak and dagger" operators. The Hermitian operator, "†", called a dagger, which changes an *s-ket* into an *i-bra*, or an *s-bra* into an *i-ket*, is a temporal conjugate transpose, $a - ib$ to $a + ib$, or the transpose of a matrix, which flips a matrix over its diagonal, when the arrow of time reverses. The application of a negative dagger "−†" changes an *i-ket* or an *i-bra* to its reciprocal. Let us now define the antiHermitian operator, "‡", called Cephas' cloak, which changes an *i-bra* into an *s-ket*, or an *i-ket* into an *s-bra*, that is, into their spatial conjugates when the arrow of space reverses, is a disjoint transpose, $-a \pm ib$ to $a \pm ib$, or the antitranspose of a matrix which turns the diagonal of a square matrix clockwise or anticlockwise to be its antidiagonal. The application of the negative antiHermitian operator "−‡" changes an *s-ket* or an *s-bra* to its reciprocal.

Hence, in the probabilistic context above, a retarded wavelet and an advanced wavelet are being projected and superimposed at a location in space-time along an axis of propagation.

The information inside the maximally entangled "black hole - white hole" pair may be expressed as the density matrix of the entanglement as follows:

$$\rho_{BH\cdot WH} = Trace_{BH\cdot WH} |\rho_{BH}\rangle\langle\rho_{WH}| \quad (8.5)$$

The entropy of the entanglement of the "black hole - white hole" pair is given by

$$S_{BH\cdot WH} = -Trace\left[\rho_{BH\cdot WH} \ln \rho_{BH\cdot WH}\right] \quad (8.6)$$

Since a black hole outside its event horizon travels toward the future, the matter and space-time gradually accreted by a black hole follow the arrow of time of the retarded wavelet and the arrow of entropy, but the white hole outside of its event ektropí follows the arrow of time of the advanced wavelet and the arrow of ordentropy, since all matter and space-time inside the white hole would eventually be gradually dispelled to its spatiotemporal medium and origin.

The Schwarzschild metric, a spatiotemporal distance function, for the General Theory of Relativity in an expanding spatiotemporal medium is an exact solution to the Einstein Field Equations that may be expressed in spherical coordinates (t, r, θ, ϕ). The Schwarzschild metric describes the curved spacetime around a black hole singularity, and in a fundamental and effective way, it provides a practical approximation to the spatiotemporal medium in the vicinity of gravitating bodies such as a planet or a star.

$$2\frac{gr^2}{rc^2} \equiv 2\frac{gr}{c^2} \qquad (8.7)$$

$$e^{-2\frac{gr}{c^2}} \approx 1 - 2\frac{gr}{c^2} + \cdots \qquad (8.8)$$

$$d\Omega^2 = d\theta^2 + \sin^2\theta \, d\phi^2 \qquad (8.9)$$

$$ds^2 = -e^{\mp 2\frac{gr}{c^2}} dt^2 + e^{\pm 2\frac{gr}{c^2}} dr^2 + r^2 d\Omega^2 \qquad (8.10)$$

The above reversible Schwarzschild metric describes a curved spatiotemporal medium that may be expanding or contracting around a black hole singularity where time and space are reciprocal, and the arrow of time may reverse itself, such as in the vicinity of a "black hole - white hole" pair.

Eternal black holes have a white hole region in their past in addition to a black hole region in the future, as a solution to the Einstein Field Equations. However, this solution is not considered to exist for black holes created through gravitational collapse. (Oppenheimer et Al, 1939) On the other hand, a supermassive black hole at the center of a galaxy may spawn a supermassive white hole. (Hawking and Penrose, 1996)

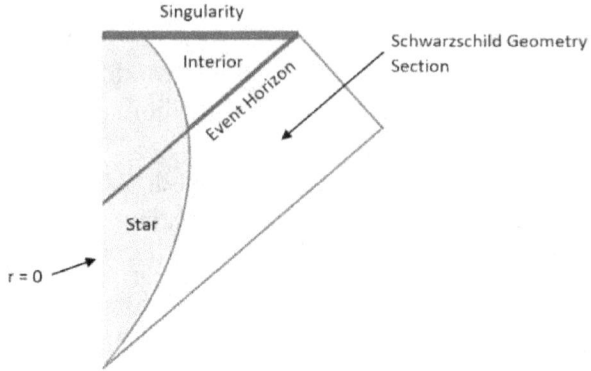

Figure 2. The Geometry of a Collapsed Evaporating Black Hole. (Penrose, 1963)

8.1. The Hawking radiation theory.

There may be hypothetical particles around the boundary of a black hole that may be described as Hawking radiation. The relativistic quantum effects outside of the event horizon of a black hole may release black body radiation also known as Hawking radiation. Local apparent horizons or event horizons are contained in the Lorentzian geometries of space-time. The kinematic effect produced by virtue of Lorentzian geometries is the Hawking radiation. The Hawking radiation suggests that the temperature of black holes are conversely proportional to their mass, so, the smaller the black hole volume, the higher its temperature and luminosity.

Hawking radiation is only theorized at the moment, but if it exists, the information that enters the black hole was theorized to be recovered from the black hole via the Hawking radiation. Hence, the information was not really lost. If Hawking radiation were observable, there would be proof that black holes abide by the laws of thermodynamics. The Hawking radiation is theorized to consist mostly of photons. Most of the Hawking radiation comes from a very large spatiotemporal region around the black hole, not from the fringe of the event horizon boundary.

It is hypothesized that the issue with information loss comes from the loss of black hole mass through the Hawking radiation while the information lost does not come back to the accessible region of the universe. So, that in the end, the black hole evaporates with the information it consumed, breaching the rules of quantum mechanics. The Hawking radiation is theorized to consist of very energetic particles, antiparticles, and gamma rays. All the Hawking radiation as well as the black holes area are invisible. The Hawking radiation may also come from a theoretical white hole, equal to the Hawking temperature of a black hole, would also be invisible from the geometry of the outside region of the white hole. Hawking radiation is theorized to decrease the rotational energy and mass of a black hole, causing the gradual evaporation of the black hole. As previously stated, the Hawking radiation is theorized to be conversely proportional to the mass of the black hole, so extremely small black holes are hypothesized to emit a larger amount of Hawking radiation than very massive black holes with faster dissipation. The time that the internal temporal volume and mass of a black hole may take to disperse is in the millions or billions of years.

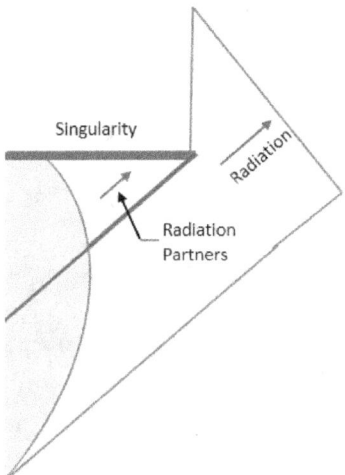

Figure 3. The Escaping Radiation and Infalling Radiation Partners of a Collapsed Evaporating Black Hole. (Penrose, 1963)

The escaping radiation is entangled with infalling radiation partners, since the interior is not measured, it is conceptualized that there might be a large entropy from the infalling radiation partners. The outside of the black hole begins as a genuine state and develops as a mixed state. The evolution of gravitation may describe how the radiation outside of the black hole seems to be in a mixed state due to its entanglement inside the black hole. A sibling universe splits from the parent universe. The interior of a collapsed black hole may be ordinary (charged with information) or non-ordinary.

$$S_{Classical} \equiv S_{Quantum\ System\ -\ External\ View \atop Ordinary\ or\ Non-ordinary} + S_{Quantum\ Fields\ Entropy} \qquad (8.11)$$

§ 9. The Lambda CDM Model as the observed amount of photon sources in the universe.

The "Λ" CDM model accounts for light elements in the universe or the amount of observable photon sources in the universe, the accelerated spatiotemporal expansion, the cosmic radiation background, and the large structure of the universe. A core component of the "Λ" CDM model also implies that the General Theory of Relativity is assumed to be correct. The scale factor "a" is related to the Hubble parameter, $H = \dot{a}/a$.

The Friedman equation is given by

$$H^2 = \frac{8\pi G}{3}\rho - \frac{k}{a^2} + \frac{\Lambda}{3} \qquad (9.1)$$

These are the components that play a role in how space-time behaves as an expanding or contracting medium: "ρ" is the energy density of the matter and radiation of the universe or $8\pi G\rho$, "k" is the cosmological curvature of the large structure of the universe, which is at this time, approximately "0", and "Λ" is the cosmological constant. The Hubble parameter is how space-time converges or diverges. If "Λ" is positive, space-time diverges in the direction of the retarded wave or the negative temporal direction (the future), if "Λ" is negative, space-time converges in the direction of the advanced wave or the positive temporal direction (the past). Time emerges to create more space, space diverges to endow more time.

Divergent or Convergent Space ≡ Past or Future Time (9.2)

$$Time \equiv Imaginary\ Space \tag{9.3}$$

$$t \equiv ir \tag{9.4}$$

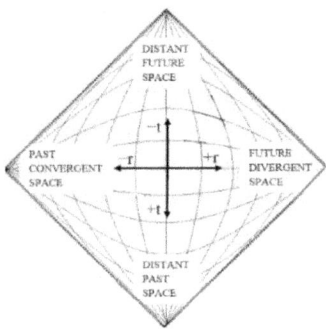

Figure 4. The Temporal and Spatial Directions of the Divergent or Convergent Spatiotemporal Medium. (Penrose, 1963)

Therefore, space and time are effectively interchangeable, and fundamentally the same thing, $t = ir = r\angle 90^0$, and $i = t/r = 1/c$, the space-time duality is an effect which becomes much more noticeable at relativistic speeds approaching the speed of light. Time may be conceptualized as a phase transition of space. So, a photon is fundamental while it also embodies the space-time duality.

Let us solve for "Λ" for the observable amount of photon sources in the universe.

$$\Lambda = 3\left(\frac{\dot{a}}{a}\right)^2 - 8\pi\left(\frac{\ddot{a}c^2}{a}\right) - \frac{3k}{a^2} \tag{9.5}$$

Therefore, the total number of photons "N" emitted from a black hole, and other radiation "γ" emitted from a black hole, during its lifetime are part of "Λ".

$$\Lambda_{BH} = -8\pi\left(\frac{\ddot{a}c^2}{a}\right) = -8\pi\hbar\left(\omega_p^2\right)N - 8\pi\hbar\left(\omega_p^2\right)\gamma = -8\pi\hbar\left(\omega_p^2\right)[N+\gamma] \tag{9.6}$$

The Page time represents the maximum von Neumann entropy, or fine grained entropy, of the Hawking radiation outside a black hole, $S = -Tr[\rho \ln \rho]$, the time when the entanglement entropy transitions from increasing to decreasing. The total number of photons "N" are emitted from a black hole during its lifetime. The "$N/2$" photons are emitted in the early period of a young black hole. These early photons are entangled with the "$N/2$" photons in the late period of an old black hole. The entropy of the encircling Hawking radiation increases for a young black hole, since an observer outside the event horizon of the black hole does not know that a photon outside the event horizon has a partner hidden inside the black hole. (Page, 2013)

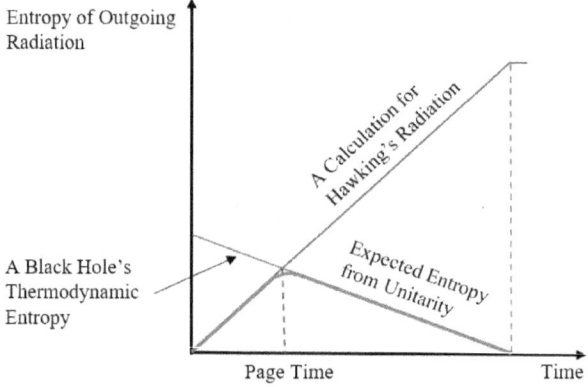

Figure 5. An Illustration of the Page Time.

Every Hawking photon returns to the old black hole to find its partner, and as a result, the total entropy of the Hawking radiation decreases. The absolute temperature of the black hole increases during the process of evaporation of the black hole, regardless of whether the thermodynamic entropy of the black hole increases or decreases during the process. The absolute temperature "T" of a Hawking photon is proportional to the mass of the black hole, "$1/M$", and the average energy of the Hawking photon "$\hbar(\omega_{avg.})$", is proportional to the absolute temperature "T", where "ω_{avg}" is the average angular frequency.

Let us posit the case of the behavior of a black hole when $M \gg M_p$, where M_p is the Planck mass. If the behavior is contentious, it is possible to suggest that a complete quantum theory like "A Dynamic Theory of Space-Time" can concerned itself effectively with the behavior.

When the total number of Hawking photons reaches "N" as the black hole mass contracts to "$M_{Contracted}$", we obtain

$$\frac{N}{2} = \int_{M_p}^{\frac{M_{Initial}}{\sqrt{2}}} \frac{dM_{BH}}{\hbar(\omega_{avg.})} = \int_{\frac{M_{Initial}}{\sqrt{2}}}^{M_{Initial}} \frac{dM_{BH}}{\hbar(\omega_{avg.})} \qquad (9.7)$$

Where "$M_{Initial}$" denotes the initial mass of the black hole. The Page time is given by the contracted mass equal to the effective initial mass, $M_{Contracted} = M_{Initial}/\sqrt{2}$. It is also theorized that a black hole may emit massive particles and gravitons. Hence, it is reasonable to hypothesize that the Page time is probably smaller than what has been denoted above, which begs a detailed mathematical analysis that may support or not support this hypothesis.

Chapter 5

The Holographic Principle.

§ 1. The ER ≡ EPR conjecture.

The conjecture "ER ≡ EPR" in physics states that two entangled particles, or two entangled black holes, are connected by a spatiotemporal bridge that may be a basis for unifying Quantum Mechanics and the Einstein-Cartan Theory of Gravity or the General Theory of Relativity, "QM ≡ ECTG", to account for curvature with torsion, or "QM ≡ GTR" to account for curvature without torsion, into a quantum gravity theory that describes gravity in the range of conditions and with the qualitative behavior, where quantum effects cannot be dismissed, to arrive at a complete and consistent quantum gravitational theory such as "A Dynamic Theory of Space-Time", a precise quantum version of GTR. (Nieves, 2020)

The most basic attribute of space-time is its connectivity, entanglement is a form of connectivity from QM. The divergence or convergence of space-time is another basic attribute that connects the GRT to the tension tendency of quantum states to increase complexity in the medium. So, wherever and whenever there is Gravitation in the spatiotemporal medium, there will also be QM. It is reasonable to formulate the correspondence dyads, $\langle GRT|QM\rangle \equiv \langle QM|GRT\rangle$, since the correspondence of Complexity in Quantum Mechanics connects to the Classical Structures of the General Theory of Relativity, and vice versa.

Is it possible that two maximally entangled black holes would function as an Einstein-Rosen bridge between their two widely separated non-interacting locations in distinct space-times, or in two very far locations, which may be expanding, in the same space-time? Would the insight, ER ≡ EPR, be physically possible?

Let us posit that if possible, this insight may provide physics with a way to reconcile Quantum Mechanics and the General Theory of Relativity. Entanglement, a nonlocal connectivity, is an attribute of inherently probabilistic Quantum Mechanics. What would be their maximal entanglement mechanism?

There might be different types of entangled black holes or "black holes - white holes" pairs. Two or more nonconcentric black holes may contain some particles that are maximally entangled, but the black holes may not be maximally entangled in a particular entangled state. Moreover, nonconcentric BWP may have magnetic field loops if the BWP co-exists on the same spatiotemporal volume. Two concentric, or nonconcentric black holes may be maximally entangled, if their particles are all entangled. Two or more distinct black holes may be minimally entangled, if some of their particles are entangled. Similarly, a black hole - white hole pair may be minimally entangled (nonconcentric supermassive BWP), or maximally entangled (concentric eternal black holes) in a particular entangled state if the black hole travels on the retarded wave of its world volume and the white hole travels on the advanced wave of the same world volume.

It is interesting to hypothesize that if each of the two black holes have particles that are maximally entangled, as the maximally entangled radiation spreads in each black hole, there may be quantum spatiotemporal bridges between the radiation of both black holes to facilitate, motivate, and manipulate the emergence of an Einstein-Rosen bridge or throat, between the two black holes. The emergence of a maximally entangled spatiotemporal bridge begs the questions of whether the spatiotemporal bridge is traversable from either side by information, or if the bridge may develop one of its black holes through an event horizon at the point of entanglement, spatiotemporal pressure differential, curvature, torsion, gravitation, electromagnetism, maximal entanglement, or other physical mechanism or agency, under the laws of transmutation and conservation of energy, that transforms low thermodynamic entropy (potential energy) into information, and/or into a white hole, without violating the laws of nature.

Furthermore, the conjecture of an event horizon at the point of entanglement between two or more maximally entangled black holes provides a mechanism for the reversal of the arrow of ordentropy to the arrow of entropy inside the event horizon of a white hole, or event ektropí, in a BWP, producing, but not limited to, an expanding spatiotemporal medium, antigravitation, and a porous membrane or horizon at the event ektropí of the white hole at the point of entanglement or beyond. Consequently, matter, energy, and space-time may pass through the membrane between the entangled black holes. Matter waves are an instance of the wave–particle duality. All matter exhibits wave-like behavior.

At the point of entanglement horizon or membrane, it is theorized a black hole advanced matter wavelet and/or a black hole retarded matter wavelet are chiral, so they reverse as each matter wavelet crosses over. Advanced and retarded matter wavelets share odd parity, a property that changes their sign under their spatiotemporal Lorentz transformation, and helicity, a translation property which represents the potential for helical flow to form. Helicity is proportional to the strength of the flow and its vorticity. The contracting spatiotemporal wavelets would expand and the expanding spatiotemporal wavelets would contract.

The spatiotemporal pressure differential from the higher pressure black hole to the lower spatiotemporal pressure black hole may initiate the cross-over process into the lower spatiotemporal pressure black hole, so that a retarded spatiotemporal wavelet crosses over, reverses, and expands, starting the spatiotemporal wave conversion process to a white hole. Any matter, energy, and space-time that passes through the point of entanglement membrane to the white hole side would be in an increasing realm of an expanding spatiotemporal medium and a decreasing realm of a contracting spatiotemporal medium until the pressure slowly stabilizes in the BWP and the cross-over process is minimized.

Aside, as the arrow of time reverses inside the event horizon of a black hole, energy decreases by a factor of "i/c^2", ($E = mc^2 \rightarrow E = im/c^2$), as the spatiotemporal medium contracts, and space and time exchange roles, on the advanced spatiotemporal wave. The factor "$1/c$" is equivalent to "t_p/l_p", which represents the reciprocal of the velocity, or the rate of change of temporal duration with respect to spatial distance, for an object traveling in the reverse direction of time.

It is proposed that a traversable spatiotemporal bridge (TSB) is equivalent to a pair of maximally entangled black holes that share quantum information that may have been safely manipulated, by gravitation-antigravitation conversion, matter-antimatter transmutation, and electromagnetism, to provide a stable traversable spatiotemporal bridge for information between two locations in space-time that may now be very far apart. The quantum information may allow teleportation of life or material objects into, through the bridge, or out of the spatiotemporal bridge, the entangled black holes that may become a black hole - white hole pair, BWP.

Could the conjecture "TSB ≡ BWP" be technologically achievable? Could a narrow and encoded counter gravitational beam from a beam emitter on a drone ship approaching $c/\sqrt{3}$ orbiting around the inside, or the outside, of the event horizon of a black hole, and/or an encoded high energy antiparticle beam from an anti-matter decelerator that causes the gradual disintegration of a singularity, gradually transform a black hole into a white hole to create a BWP?

The circuitry of entanglement of two or more spatiotemporal bridges are temporally oriented from start to finish, so that the correlated entanglement quantum signals of information travel through the spatiotemporal bridges of the word lines or the world volumes from the present time to the point of entanglement between particles, or black holes, in the past, then forward to the future through the world lines or the world volumes, arriving instantaneously as measured by a clock in the present near each entangled particle or black hole. Thus, entanglement can occur in the present space, but it is maintain through the past world lines of particles or the past world volumes of black holes.

The present or future spatiotemporal medium of entanglement would be smooth and uncluttered of the circuitry web of entanglement. A tachyonic particle, or a white hole, would travel its own past course on the advanced wave through the entanglement web of the past. Even though space-time is emergent and fundamental, the spatiotemporal medium is the background where the quantum entanglement interconnections between particles or black holes exist. Entanglement is the architecture of an entangled black hole pair, and teleportation is its method of transportation.

When entanglement occurs between two particles or black holes, advanced and retarded wavelets of quantum entanglement emerge carrying, or teleporting, quantum information between the entangled objects. The advanced and retarded wavelets are entangled and share quantum states, so it is theorized that the wavelets handshake and exchange quantum information with each other, as long as they share the same quantum states, and there is an update. Each packet of quantum information has the previous quantum states to handshake and the new quantum states to entangle. The quantum entanglement wavelets are radial resultants of the chosen coordinate system by an observer from the point of entanglement to the point where each entangled object is located in space-time at any given instant of time.

Aside, it is possible to hypothesize that a holographic reality may be expressed as a (3 + 3) formalism which has been counter projected from a lower dimensional advanced information repository or memory, from within a parallel system of a concentric reality. In that context, the events of information, or the best ideas of information, come from all there was, is, or will be, at a perfect time since the mechanism, with its memory of all events, stands within the existence that may control the temporal velocity or acceleration, as well as the background themes, of the holographic medium for optimization and improvement. Who would be the sovereign and/or the all-knowing existence, of such an advanced and extensive complexity? What would be its constant notions, purposes, and goals for each possible event or entity? Would the sovereign existence dwell in a parallel universal reality, very high above the frequency of a holographic universe? Life seeks its duality and true origin, even if the origin of life seems to be metaphysical. Is the essence of life also a holographic-metaholographic duality? (NSAB, 1995)

Infinity does not need a beginning or an end, only a just purpose and very reliable plans. All there is in our universe may be entangled to some degree; entanglement is essential to existence. It may be the proper time to ask the following rhetorical questions: Does the consciousness of a divine existence, or the consciousness or the metaconsciousness of a lifeform, utilize a similar multidimensional holographic principle, and/or a quantum spatiotemporal bridge, for entanglement? Would the retarded infinity and the advanced infinity meet at a continuity point of holographic projection in the very remote past or the very far future in a universal cycle of existence?

§ 2. The principles of entanglement for all there is inside the event horizons of black holes.

The holographic principle, for the entanglement of two or more black holes, theorizes the encoding of information projected onto a surface area from a multidimensional reality. The hologram may illustrate a complex representation of reality inside the event horizon of entangled black holes.

1. The complex spatiotemporal medium is a principle of coherence between all that is spatial and all that is temporal in nature. The temporal wavelets that coexist above the spatial wavelets of a volume will interfere, entangle, and merge, creating a coherent spatiotemporal reality. All that exists in the temporal volume at the speed of light is constant or time invariant.

$$\lambda_p = \frac{-\hbar c}{\frac{\partial C_p}{\partial \tau}} = \frac{-\hbar c}{-m_p c^2} = l_p \qquad (2.1)$$

$$T_{Period} = \frac{1}{f_p} = \frac{\hbar}{\frac{\partial C_p}{\partial \tau}} = t_p \qquad (2.2)$$

$$c = \frac{l_p}{t_p} \qquad (2.3)$$

2. The Planck frequency and the wavelength of an instant of an entangled object, of all there is, are proportional. Every string of reality is characterized by its frequency and wavelength. All spatial and temporal entangled objects, particles, or media may have a characteristic frequency. Any object, particle, or medium, may have a unique frequency and wavelength as the object travels through space, but if the object, particle, or medium, does not travel through time, the object, particle, or medium may be tuned to the frequency, or frequencies, and wavelength, or wavelengths, of its temporal wavelet and the harmonics of the medium, with other objects or particles. If the object, particle, or medium, travels through time or expands, converges, or diverges temporally, the object, particle, or medium may tune to a distinct or corresponding frequency, or

frequencies, or to a wavelength, or wavelengths, of its temporal wavelet or the harmonics of the medium, with other objects or particles. Space endows Time, and Time endows more Space. Motion endows Change, and Change allows measurement of Space and Time.

3. For four-dimensional energy or power, or Einsteinian energy or power, we have

$$\hbar(\omega_p) = m_p c^2 \qquad (2.4)$$

$$P_{avg} = \frac{1}{T}\int_0^T p(t)dt \triangleq \frac{1}{2\pi}\frac{dW}{dt} \qquad (2.5)$$

$$\frac{m_p c^2}{2\pi t_p} = \frac{\hbar(\omega_p)}{t_p} = \hbar(\omega_p^2) \qquad (2.6)$$

2.1. Photon production from the spatiotemporal field.

The spatiotemporal wave field, or space-time field, is like a collection of spatiotemporal oscillators, each wavelet oscillator has its unique angular frequency. If the wavelet oscillators are at the minimal energy state, the spatiotemporal field is at the ground energy state or at the time dependent minimal volumetric state. If a spatiotemporal wavelet is excited above the ground energy state to its first quantum energy state, a single photon of a unique angular frequency would be manifested surrounded by the existing ground energy state of the spatiotemporal field. Therefore, a photon is generated by a spatiotemporal phonon, or two photons may be created by the collision of two oscillating spatiotemporal wavelets.

$$\hbar\omega_p \equiv \hbar\upsilon \qquad (2.7)$$

$$\hbar(\omega_p^2) \equiv 2\hbar\upsilon \qquad (2.8)$$

1. If a particle and a corresponding antiparticle collide, annihilation occurs, and their mass is converted into radiation energy,

producing two photons in the process, not a single photon because one photon would remove momentum, which is not allowed without any external forces acting. It is reasonable to state that the virtual phonon may be more fundamental than the photon. The phonon can produce light as well as heat, and space-time is emergent and fundamental. Spatiotemporal divergence or convergence is pair production or annihilation depending on the arrow of time and the angular frequency, due to the elongation or contraction of spatiotemporal wavelets at extremely high angular frequencies such as those predicted for the nearly constant and very rapid hypothetical cosmic inflation period.

Is it possible that at the beginning of the universe, three-dimensional time emerged as a resultant, a folded proper time, an oscillating and translating scalar field, close to a "de Sitter" formalism, to unfold space and time three-dimensionally, or multidimensionally, while temporal fluctuations emerge as temperature changes, spatial fluctuations emerge as pressure or energy density changes, increasing complexity and causing spatiotemporal regions of expansion or contraction, in the universal formation of galaxies, stars, planets, moons, and other divine creations?

However, the presently unobserved creation of electromagnetic waves or gravitational waves may occur at extremely low angular frequencies. Any perturbation that initially existed in the spatiotemporal field was extremely magnified, longer than the diameter of the observable universe, producing a relatively smooth and immensely uniform universe at very large scales.

An example of the quantum realm influencing the classical realm of the universe. The divine design of a universe, with very high temperature, includes the generation of, but not necessarily limited to, particles, spatiotemporal waves, matter, and radiation energy, for the formation of all there was, is, and will be. The creation of celestial bodies, nebulas, and galaxies does not necessarily require an immensely uniform density as observed in the cosmic microwave radiation background due to gravitation.

The Planck mass of a Planck string is a function of angular frequency. If the Planck angular frequency of a Planck mass increases to an extremely high level, the oscillating Planck mass of a string would stretch to its maximum amplitude, where the amplitude may be theorized to be about $\sqrt{3}$ times the spatial Planck length, $m_p \to \left(\sqrt{3}\right) l_p$.

$$\lim_{\omega_p \to \omega_p^2} m_p^2\left(\omega_p^2\right) = R_S^2 - l_p^2 \qquad (2.9)$$

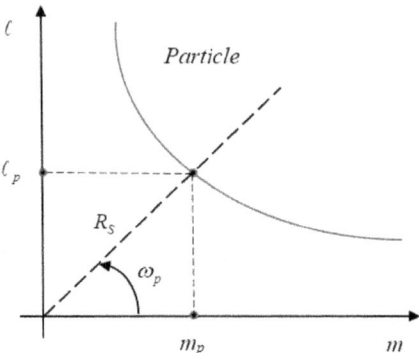

Figure 1. The Compton wavelength of a particle of mass "m" is represented by a hyperbolic curve. The intersection between the straight line and the curve defines the Planck mass, m_p, and the Planck length, ℓ_p, at the Planck angular frequency, ω_p. The straight line is the linear increase of the Schwarzschild radius, $R_S = 2GM/c^2$, with respect to a particle mass in an entangled black hole.

The mass of a photon is relativistic, as the photon energy travels, or it oscillates, through space, not through time. Consequently, the Planck mass may transmute into light, or into a spatiotemporal ratio "c", Energy can turn into Mass, Mass can turn into Light, and Light is pure Energy.

$$\hbar\left(\omega_p^2\right) = m_p\left(\omega_p\right)c^2 = m_p\left(\omega_p\right) \cdot \frac{l_p^2}{t_p^2} = \frac{\sqrt{R_S^2 - l_p^2}}{t_p} \cdot \frac{l_p^2}{t_p^2} \qquad (2.10)$$

$$\hbar(\omega_p^2) = \frac{(R_S^2 - l_p^2)}{t_p} \cdot c^2 = \frac{(\sqrt{3})l_p}{t_p} \cdot c^2 = (\sqrt{3}) \cdot c^3 \quad (2.11)$$

The four-dimensional cube of light times the resultant holographic coefficient "$\sqrt{3}$" of the relativistic mass value, $m_p \triangleq (\sqrt{3})l_p$, represents a spatiotemporal tesseract with three temporal dimensions folded for the illustration of a (3 + 1) formalism, where $|l_p| \gg |t_p|$.

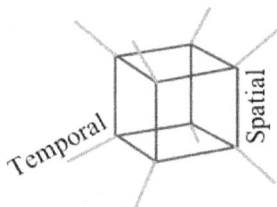

Figure 2. A Four-Dimensional Cube of Light with Folded Temporal Dimensions.

For six-dimensional energy, or Hawking-Feynman energy, we obtain

$$\hbar(\omega_p^2 c) = m_p(\omega_p)c^3 = m_p(\omega_p) \cdot \frac{l_p^3}{t_p^3} = \frac{(R_S^2 - l_p^2) \cdot l_p^3}{t_p^4} = \frac{(\sqrt{3})l_p}{t_p} \cdot \frac{l_p^3}{t_p^3} = (\sqrt{3})c^4 \quad (2.12)$$

The six-dimensional tesseract, or hypercube, of light illustrates a holographic spatial cube within a holographic temporal cube for a (3 + 3) formalism. It is theorized that the (3 + 3) formalism is a fully extended representation of the six-dimensionality of space-time. The product of the resultant holographic coefficient "$\sqrt{3}$" and the resultant spatiotemporal expansion "c" beyond the spatial cube, since the six-dimensional Hamiltonian energy is "c" times greater than the four-dimensional Hamiltonian energy, $dH_6 \equiv c \cdot \delta H_4$.

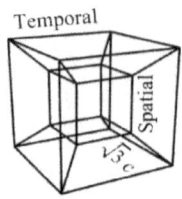

Figure 3. A Six-Dimensional Tesseract of Light with Unfolded Temporal Dimensions.

Aside, all other things being equal, the body or form of a lifeform that could exist in such a theoretical reality of extremely high Planck angular frequency would have a body of light and may be virtually weightless in a planetary gravitational field. If such a lifeform existed, it would be unseen by human eyes, and the thinking process of that lifeform would be incredibly faster than in a more dilated time at a lower frequency, or in a stronger gravitational field. The six-dimensional energy equation illustrates how mass may transmute into light, or electromagnetic energy, proportional to the square of the Planck angular frequency, ~ 3.44×10^{86} rads/sec, and the spatiotemporal dimensions. (Nieves, 2020)

2. The existence of a retarded wavelet and an advanced wavelet in complex space-time affirms the principle of duality or compossibility. A duality in this context is a hidden equivalence between two seemingly different wave functions. Charge, spatiotemporal wavelets, matter, temperature, etc., in nature, are entangled duals that may coexist depending on the direction of the arrow of time. For example, for the relativistic mass value, the holographic coefficient becomes, $\pm m_p \triangleq \left(\sqrt{\pm 3}\right) l_p$, depending if the spatiotemporal medium is converging or diverging. If the tesseract's volume diverges, $m_p \triangleq \left(\sqrt{3}\right) l_p$, and if the volume converges, $-m_p \triangleq \left(\sqrt{-3} l_p\right) \triangleq i\left(\sqrt{3}\right) l_p$.

3. The principle of oscillation is the foundation of the motion of spatiotemporal wavelets. Spatiotemporal wavelet oscillation may be a product of Planck angular frequency, temporal emergence may endow Planck angular frequency and motion from all there is.

Planck quantum complexity inside the event horizon of entangled black holes is directly proportional to the amplitude of spatiotemporal wavelet oscillations of the medium. The handedness of oscillation determines the direction of propagation of spatiotemporal wavelets, including the arrow of time within a specific spatiotemporal medium. Wavelets are Oscillations, Oscillation is Motion, and Motion is Change.

As the old adage, with a new outlook, states: the only thing constant in oscillation, at less than the speed of light, is change. So, let us examine the constancy in the ratio of changes in absolute temperature per angular frequency for an entangled black hole.

$$\frac{T_p}{\hbar(\omega_p)} = \frac{1}{k_B} \tag{2.13}$$

$$\frac{T_p}{\omega_p} = \frac{\hbar}{k_B} \tag{2.14}$$

$$\frac{\partial T}{\partial \omega} \equiv \frac{\hbar}{k_B} e^{\frac{\hbar}{k_B}} \equiv \frac{T_p}{\omega_p} e^{\frac{T_p}{\omega_p}} \quad T \geq T_p \text{ and } \omega \geq \omega_p \tag{2.15}$$

Therefore, as absolute temperature rises from the Planck temperature, oscillations (Planck phonon wavelets) increase proportionally to $\hbar/k_B$, ~ 7.638232578 × 10^{-12} K/rads/sec. Consequently, black hole angular frequency increases from the Planck angular frequency within the event horizon of an entangled black hole, and the Planck quantum complexity rises in the direction of the arrow of time or the arrow of thermodynamics during formation.

Therefore, it is possible to theorize that the amplitude of the oscillations are directly related to the kinetic energy of the phonons of spatiotemporal quantum complexity. The spatiotemporal quantum complexity has been defined as the sum of the amplitudes of spatiotemporal wavelets inside the event horizon of an entangled black hole.

The absolute temperature of a phonon inside the event horizon of a black hole, as the virtual mass of the kinetic energy of a phonon would stretch to its maximum amplitude, or if the virtual mass oscillates at an extremely high frequency, the amplitude may be theorized to be $\sqrt{3}$ times the spatial Planck length, $m_{Phonon} \to (\sqrt{3}) l_p$, given by

$$T_{Phonon} = \frac{dE_{Phonon}}{c \cdot dS_{Phonon}} \quad (2.16)$$

$$T_{Phonon} \triangleq \frac{1}{2} m_{Phonon} c^2 \equiv \frac{\sqrt{3}}{2} \cdot l_p \cdot \frac{l_p^2}{t_p^2} \quad (2.17)$$

Where dE_{Phonon} is the six-dimensional energy of a phonon, and dS_{Phonon} is the four-dimensional entanglement entropy of a phonon during formation of an entangled black hole. (Nieves, 2020)

Thus, the absolute temperature of an entangled black hole during formation may be denoted in terms of the Planck quantum complexity as

$$T_{BH} = \frac{C_p}{2^N} \cdot T_{Phonon} = \frac{\sqrt{3}}{2} \left(\frac{C_p}{2^N} \cdot l_p \cdot \frac{l_p^2}{t_p^2} \right) \quad (2.18)$$

$$T_{BH} \equiv \frac{\sqrt{3}}{2} \left(\frac{C_p \cdot l_p}{2^N} \right) \cdot c^2 \equiv \frac{\sqrt{3}}{2^{N+1}} \left(C_p \cdot a_{C_p} \right) \quad (2.19)$$

The absolute temperature inside the event horizon of a black hole is T_{BH}, and a_{C_p} is the volumetric acceleration of the spatiotemporal medium inside the event horizon of an entangled black hole, during formation.

4. The principle of causality assures that every spatiotemporal event

may be the effect of the purpose of a cause, not a coincidence. Nonetheless, quantum entanglement breaks causality. Every event is a place and time on the world volume of an entangled object. It is always possible to use transcendental numbers to represent randomly distributed numbers not connected to this principle.

5. The principle of complementarity describes the attributes of complement objects in nature. An entangled object may complement another like object, with opposite but equal attribute, such as handedness, spin, polarity, helicity, charge, etc., to achieve the cause or the effect at an event on its world volume. The wave and particle theories of light may be able to explain a set of phenomena, although each theory separately only accounts for some aspects of the phenomena. The absolute energy of the combined action of complements is fundamental for the pair creation process.

Chapter 6

Will Our Universe Diverge and Then Converge like an Evaporating Black Hole?

§ 1. The universe as a black hole - white hole pair.

The nuclear force that keeps ordinary atoms from collapsing is not infinitely great. The Chandrasekhar limit is the maximum mass of a stable white dwarf star. The currently accepted value of the Chandrasekhar limit is about 1.4 M☉ (2.765×10^{30} Kg). The limiting value for neutron star mass, analogous to the Chandrasekhar limit, is known as the Tolman–Oppenheimer–Volkoff limit. White dwarfs resist gravitational collapse primarily through electron degeneracy pressure, compared to main sequence stars, which resist collapse through thermal pressure. Once collapse begins, not all atoms hold up mass endlessly piled on their mass. The nuclear force of ordinary atoms cannot withstand an intense gravitational pressure to collapse the nucleus and the electrons. If the core of a star has a mass under three times that of the sun, neutron pressure protects it from complete collapse leading to the creation of a neutron star. Otherwise, a massive star can collapse into a stellar black hole. It is now an opportune time to ask the following rhetorical question.

What if the universe is the result of a connected traversable supermassive black hole - white hole pair?

If the black hole in the black hole - white hole pair swallows the white hole, the resulting spatiotemporal geometry becomes a continuum where "all" mass, matter, and energy, coming out of the white hole falls into the black hole in a continuous cycle of expansion and contraction. The object falling into the black hole toward the singularity is contracted toward "$-\omega$," through the throat of entanglement, only to expand after a big bang into its conjugate white hole with diverging spatiotemporal medium, eventually falling back into its conjugate black hole with a converging spatiotemporal medium toward the singularity, continuing a cycle of universal existence. Could that represent a self-contained universe in a growing multiverse?

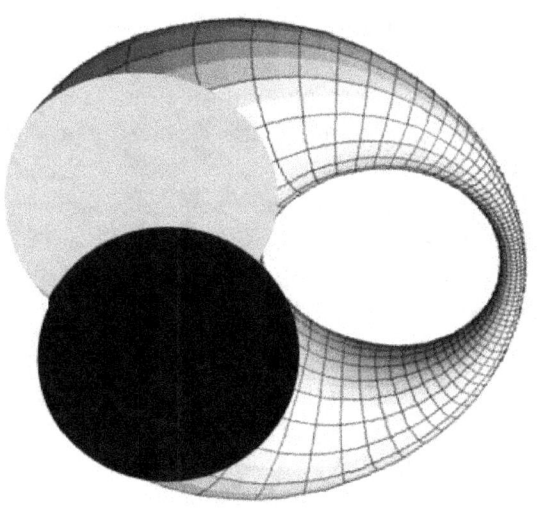

Figure 1. A connected traversable supermassive black hole - white hole pair.

1.1. The converging universe.

A collapsing star, if massive enough, can contract with such a force that even the nucleus and electrons will collapse under the immense pressure while even the nuclear force will bend its knees to the gravitational force of the collapsing star. There would not be a stopping point to the collapse. Once the nuclear force caves in, there is no stop left to withstand the enormous pressure of classical gravitation, even though gravitation is considered the weakest of the four forces of nature. The weakest force becomes the unstoppable force when the accrual of mass becomes critical. Thus, the star crumbles through all the physical barriers that may have been on the way toward negative infinity. The volume of the star keeps converging indefinitely to a point-like mass at the geometrical center while the gravitational field at the surface increases tremendously. The critical point is about three times the mass of sol, our sun. Any star, more massive than three times sol would haplessly collapse to a singularity. Moreover, if a star is more than about twenty times the mass of our sun, it cannot shed enough mass if it supernova to become a neutron star or a white dwarf, contracting instead to a singularity. So, as soon as the nuclear fuel of the collapsing start depletes, gravitational collapse would be inevitable.

1.2. On the formation of a black hole.

Stars with a mass greater than about three times the mass of our sun are bound to suffer the terminal collapse after the process begins, as those masses that are less but above the Chandrasekhar limit. What happens to the gravitation of the atoms of the collapsing star? What happens if a white dwarf collapses even more toward its geometrical center? The gravitation of the collapsing star would increase evenly as the surface area decreases proportionally toward that point-like mass, or singularity, where the contraction is heading to. The heaviest observable white dwarf has a mass of about 1.2 solar masses, while the lightest weighs only about fifteen per cent of one solar mass. Not all white dwarfs exist in isolation, and a white dwarf that is accreting material from a companion star in a binary system can give rise to several different eruptive phenomena. When a dwarf star of one solar mass collapses it has an escape velocity equal to the effective velocity of light, $c/\sqrt{2}$, so as the dwarf star collapses even more the escape velocity approaches the speed of light. The German physicist Karl Schwarzschild calculated the value of the radius of a celestial body where this phenomenon happens is called the Schwarzschild radius, $r_s = 2GM/c^2$.

$$\frac{1}{2} = \frac{GM}{R_s c^2} \qquad (1.1)$$

$$\frac{v_r^2}{c^2} \equiv \frac{GM}{rc^2} \equiv \frac{1}{2} \qquad (1.2)$$

$$c = \sqrt{2}(v_r) \qquad (1.3)$$

$$v_r = \frac{c}{\sqrt{2}} \qquad (1.4)$$

Hence, the velocity of the surface of the object of mass, falling through the boundary of the event horizon of the Schwarzschild black hole, may be at the effective value of the velocity of light, $c/\sqrt{2}$, which is about ~70% the speed of light.

1.3. The point-like mass at the center of a collapsed star is known as a Schwarzschild singularity.

The Schwarzschild radius for the collapsing solar mass of a star like our sun is about three kilometers. As a white dwarf with radius of three kilometers would have a mass density equal to thirteen times the original mass before collapse. The surface gravitation of such a dense object would be millions or billions of times the gravitation of the earth, and a car would weigh trillions of pounds, depending on the severity of the contraction.

The tidal effect of the collapsing Schwarzschild singularity would be about fourteen times that of an ordinary dwarf star undergoing a similar process. The escape velocity of such a contracted celestial object would approach the speed of light. If the contraction decreases under the Schwarzschild radius, it would resemble the convergence that happens after saturation inside a black hole, and the escape velocity would be time-like, or greater than "c" from the point of view of the retarded wave. Only a tachyonic particle would move at such a speed backward in time. So, any object of mass that falls through the Schwarzschild radius would be attracted to its singularity.

Objects of mass, matter, particles, or radiation that cross the Schwarzschild radius, or event horizon, of the collapsing object of mass, will not be able to escape the enormous pressure toward the Schwarzschild singularity at the center of collapse, or for that matter a collapsing white dwarf. Gravitation relies on pressure not attraction, things are not pulled in, they are pushed in. This phenomenon acts as a spatiotemporal hole that seems to be of infinite depth. Not even light can escape the spatiotemporal boundary such a dark star. Photons may be massless, but they have relativistic mass, as photons may lose kinetic energy in a gravitational field, experiencing gravitational redshift. Therefore, nothing escapes from a collapsing star when matter, mass of object of mass falls through the event horizon, not even light. All the objects of mass, matter, particles, will not be ejected, or come out of the hole, once they cross the event horizon of the collapsed star.

The incredibly collapsed object of mass is a hole in space-time and originally it was thought to be black, since nothing could be emitted, not even light.

Consequently, the unobserved spatiotemporal hole became known as a black hole. But very soon that name became a misnomer because the British physicist Stephen Hawking theorized in the 1970s that black holes emitted radiation from virtual particle pair production near the event horizon of the black hole.

Nevertheless, the name "black hole" captured the imagination of the academics and the public, and the name stuck. Black holes emit Hawking radiation.

There are several types of observed or theorized stable celestial objects that may not collapse:

1. Planets, Planetoids, Moons, and Asteroids – consist of fundamental particles, atoms, molecules, and matter, up to masses that are orders of magnitude bigger that any of our biggest solar planets, according to the latest astronomical observation from powerful telescopes.

2. Black Neutron Stars – Neutron stars are the smallest, densest stars that exist in our universe, they occupy a place between a conventional star and a black hole. The diameter is about twelve miles wide. The neutron stars are so dense that a teaspoon of neutron star mass would weigh about one billion tons. The neutron star material is theorized to be ten billion times stronger than steel. The surface of the neutron star is theorized to bs smooth and consists of pure neutrons. Black neutron stars are theorized to have lost a lot of their energy that they are no longer observable. The mass of a black neutron star is no more than 1.4 solar masses.

3. Black Holes – The infamous dark star may be of any value of mass, from micro black holes to supermassive black holes. There are supermassive black holes at the center of every galaxy in our universe. Black holes emit radiation, and have accretion disks that may be luminous.

4. Black Dwarf Star – A black dwarf star is what is left after a white dwarf star loses all of its energy, but retains its mass. A black dwarf star is a theoretical stellar remnant, specifically a white dwarf star that has cooled sufficiently to no longer emit significant heat or light. Black dwarf stars have masses that range up to 1.4 solar masses. It is theorized that there are no black dwarf stars that are observable yet. Our universe is actually far too young to have black dwarf stars.

So, what could happen to these supposedly stable celestial bodies? No one knows how stable these celestial bodies may be if they were isolated in space-time. What changes would they undertake if isolated from all other celestial bodies? It is a well-known fact that the only constant in the universe is change itself. Therefore, human observation, exploration, and astronomical science has not been around for as long as these celestial bodies that exist for very long periods of time, to have any records of such changes yet. Human civilization on earth is still very young and recent in terms of time scales. However, with more sophisticated and powerful telescopes reaching out into the vast of space-time and very capable super computers, scientists and researchers will learn far more about these seemingly stable celestial bodies in ways that may not be realized yet.

1.4. So, how could a celestial body change its mass?

A planet with an atmosphere may gradually lose some of its mass as the stratosphere diffuses into the surrounding spatiotemporal medium. A planet, planetoid, moon or asteroid may gain mass, thousands or millions of pounds, from the collision of particles emitted by a sun, from meteorites, comets, or matter coming from other places in a solar system, or from anywhere else in the universe over time. Nevertheless, planets are massive celestial bodies that may gain a lot more mass that they would gradually lose to the surrounding spatiotemporal medium. Consequently, a planet, planetoid, moon, or asteroid may gain more mass than it would gradually lose over its lifetime. On the other hand, a star, like our sun, may gradually lose mass through its solar flares from its fusion process when it emits nucleons or other radiation to its surrounding spatiotemporal medium.

In fact, the spatiotemporal medium is full of particles and matter, as the planet, planetoid, moon or asteroid, travels though the spatiotemporal medium. Conversely, black holes do not lose their mass like stars, planets, planetoids, moons or asteroids. However, if black hole - white hole pairs exist as theorized, then white holes would eject particles, mass or matter through its event ektropí, because as we have discussed other celestial bodies eject significant quantities of particles of mass, or amounts of matter, as they collapse or as they eject particles for internal processes, to produce measurable radiation that is observable.

It is interesting to note that the inclination of the laws of physics as well as the divine design of our universe to accrete particles, mass, matter, and energy from the large celestial bodies from the amalgamation of the miniscule celestial material of masses in the spatiotemporal medium. As star forms, during the creation of solar system, from the material that eventually may become planets, moons, asteroid belts, rings around a celestial body, and planetoids, a celestial object may accrue enough mass past its critical nuclear threshold to self-inflame as a miniature young star, that may eventually become a white dwarf star, and possibly decay into a theoretical black dwarf star.

It is interesting to speculate that the theoretical miniature young dwarf star may evolve to the condition of a stable black dwarf star while traveling through its spatiotemporal medium scooping up the particles of mass, or matter, to collapse its atomic structures into one of the smallest and densest celestial bodies that are hypothesized to exist, a neutron star, which may continue to accrete as much mass as it may encounter along its way as it travels its path through its spatiotemporal medium.

Therefore, it is reasonable to state that the infamous supermassive black holes may very well be the most stable celestial objects after all, for millions or billions of years. It has been hypothesized that during the very long , almost eternal lifetime, of a supermassive black hole, all other things being equal in the physical processes, and the courses through time, that have been discussed; the universe would end up inhabited by only celestial bodies that are stable black holes that may also amalgamate into a single megamassive black hole for all time, or until the universe starts converging like a black hole after saturation. Black holes are the existing actors on the long-term universal stage, the rest of the storyline will gradually reveal itself.

1.5. How and when were the existing black holes detected?

Cygnus "X - 1" is the first black hole to be identified by independent researchers in 1971. Astronomers detected x-rays from a bright blue star orbiting an unseen and unknown dark star. It was hypothesized that the source of the x-rays was coming from the mass, matter and energy pulled apart from the from the luminous blue star that may be accreted by a dark star, like a black hole. When massive stars collapse at the end of their lifetime they form stellar black holes. Black holes can grow after formation by accreting mass, matter, and energy from their surrounding medium. A black hole, or dark star, is a spatiotemporal volume whose gravitation is so strong that no object of mass, matter, or energy, can escape from it, not even a massless photon. John Michell, an English natural philosopher and clergyman, also, a geologist and an astronomer, first announced this remarkable idea in 1783.

1.6. How do we see or detect the invisible?

Astronomers through their powerful telescopes may identify when a near star gets too close to a black hole for the mass, matter or energy is accreted by the strong gravitation of the black hole while x-ray radiation is emitted from the cohesion. Direct evidence of Black holes have been recorded for their existence, along with observable effects of the strong gravitation. Cygnus "X - 1", in the constellation Cygnus, is part of a binary system, a star that orbits a black hole that is in its accretion process. Cygnus "X - 1" was discovered in 1964 during a rocket flight, and is one of the strongest sources X-ray sources detected from the Earth, producing a peak X-ray flux density of about 2.3×10^{-23} $W/(m^2 \cdot Hz)$. (Lewin et alia, 2006) (U.S.N.O., 2010)

The detection of black holes has been a gradual process. A hard star type to detect was the white dwarf because of their dim light and small volume, making it harder to detect that ordinary stars with more luminosity. Neutron stars were even harder to detect because they are dimmer than white dwarf stars, with little radiation of light, but with very detectable microwave pulses that made them more observable. In the case of a stellar or a supermassive black hole, it was the gravitational field from the singularity inside the black hole event horizon that provided an indirect way to detect them. It was not any light or microwave radiation coming from the black hole. In order for the gravitational field to be there, the mass of the singularity is theorized to also be there contracted and at the geometrical center as the source of the gravitational field, which also contributes the spherical symmetry of the black hole. Gravitation is the weakest of the four forces of nature; this attribute of gravitation is displayed by the black hole, because the complete gravitational field of the contracted mass is not stronger by the same amount of mass in any other physical form.

The gravitation of any star is mitigated equally regardless of distance in the spatiotemporal medium by the inverse square law, which says that the intensity of the gravitational field of a nonrelativistic mass equals the inverse of the square of the distance from the source of the field. The radiation from a point-like mass gets smaller the farther away it is from the source. So, the gravitational field of a very massive star is so mitigated at galactic distances that its gravitational effect may be almost undetectable, unless there are gravitational waves created by collision of black holes. In fact, it may be easier for an object of mass to be pushed into a black hole by the gravitational field of the singularity than for an identical object of mass to fall toward the core of an ordinary massive star. The gravitational field of the black is tremendously contracted, with a very high density, and incredibly stronger field effect, in comparison to the ordinary massive star, since the surface of the star is far from its core, and its density through its radius is a lot lower.

1.7. How could the gravitational field of a supermassive black hole be detected at a galactic distance?

One of the first ways to detect a gravitational wave is to observe how light behaves when it glides lightly over the surface of a black hole. The light ray would bend toward the source of a gravitational field. The curve of the path of the light ray is observable, detectable, and measurable. The renowned physicist Albert Einstein proposed such an effect in the early twentieth century for a ray of light gliding lightly over the surface of our sun during a solar eclipse, which was proven correct. General Relativity says that time is also stretched so the deflection is twice as great as that predicted from Special Relativity.

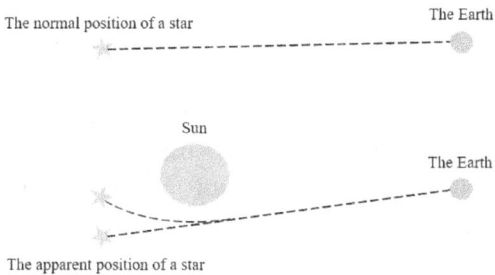

Figure 2. The Einstein's Prediction for Spatiotemporal Lensing or Warping.

It is reasonable to extrapolate that the same method may be used with a known black hole in a distant galaxy from the earth, with the light rays on all sides converging toward the earth in a gravitational lensing effect.

The image from the galaxy may be magnified in the direction of the earth producing the lensing effect. Nevertheless, if an observer were at right angles to the same black hole and galaxy, the observer would not see the same lensing effect because the light rays going to the observer may go through a spatiotemporal medium that is not magnifying the image of the same galaxy.

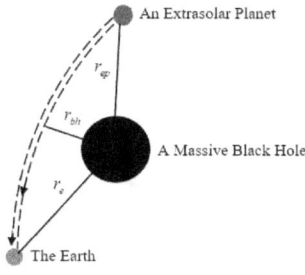

Figure 3. An illustration of the deflection of photons traveling from an exoplanet to the Earth, as it glides lightly on a spatiotemporal trajectory, outside the event horizon, near a massive black hole. (Not to scale)

Gravitational lensing has become a predominant instrument in cosmology and astronomy, having many applications. Consequently, gravitational lensing is utilized often to validate General Relativity, in the strong gravitational field scope.

Gravitational waves due to the collision of black holes have been detected by supersensitive detectors such as the LIGO detectors in Livingston, Louisiana, and in Hanford, Washington. Each detector consists of a long L-shaped structure with two arms that are 2.5 miles long (4 kilometers). A laser beam shines down each arm from the crux of the L-shape structure and mirrors at the ends of these arms reflect the light back. The physical distortions of gravitational waves in the spatiotemporal medium have been detected by these laser beams. If the beam from each arm arrives at the same time to the crux, the beams cancel each other out, and no signal is produced in the detector, which means that no gravitational wave passed through. However, if one of the beams arrives a little later than the other, that might be evidence that a gravitational wave was detected, as long as it is not due to other possible reasons that could produce that effect in the equipment.

Gravitational waves are theorized to distort the fabric of the spatiotemporal medium, and do not interact significantly with matter. The spatiotemporal length along a direction may stretch so extremely slightly along one arm of the detector and contract the other arm, about one one-thousand the diameter of a proton. This gravitational effect has now been detected at the LIGO detectors in Livington, Louisiana, and Hanford, Washington, due to the collision of two supermassive black holes.

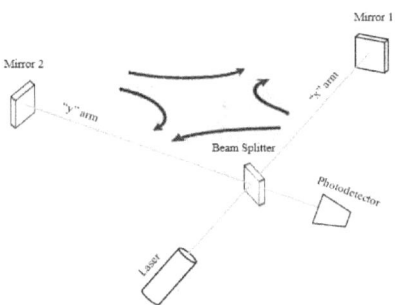

Figure 4. An Illustration of a LIGO Detector.

Black holes are infamous for accreting any matter, mass, and energy that may come near them. There are celestial bodies that orbit around black holes. Large-scale objects of mass would be fragmented and pulverized into dust and gas, as the material orbits in the accretion disk outside the Schwarzschild radius of the black hole, until it is drawn in and falls past the event horizon toward the singularity.

Matter, mass, and energy in all their forms may orbit the black hole, possibly colliding with each other while losing or transferring kinetic energy, and gradually spiraling down toward the event horizon to finally pass the point of no return, moving or spiraling radially inward on their way to be contracted and compacted as additional material for the singularity, strengthening the gravitational field of the black hole.

The light rays gliding lightly the surface of a black hole above the event horizon are observed to be bent by gravitational lenses produced by the black hole between the Earth and the black hole. It is possible to see two or more identical images of the same background black hole. In some cases, the light ray from background black holes or galaxies can be warped to form rings. Since the amount of warping depends on the mass of the foreground black hole or galaxy, you can estimate the total mass of the foreground black hole or galaxy.

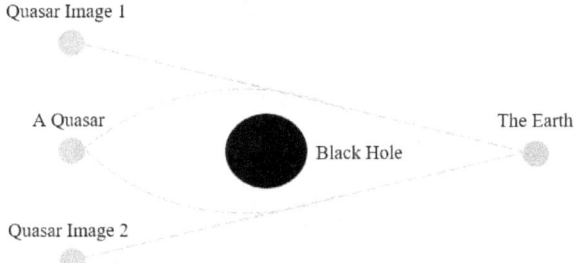

Figure 5. An Illustration of a Black Hole's Gravitational Lense.

There would be a constant influx through the event horizon into the black hole of inward spiraling particles that would lose their kinetic energy or their virtual radiation partner, transferring thermal energy, and expanding and contracting from the tidal forces that might be present. The stress and/or strain would raise the inside temperature to tremendous levels, emitting X-rays radiation in the process.

The fact that black holes accrete matter, mass and energy, make them observable, but not so easy to be observed. If the black hole has matter, mass, or energy swirling inward where there are a great number of stars at close proximity to each other where there is a lot of stellar material to get to a pitch-black black hole to form an observable accretion disk after formation. A black hole that is as black as pitch surrounded by empty space would not be very easy to observe, but if it accretes matter, mass and energy, and emits enormous X-rays, the X-rays are detectable with a powerful X-ray telescope. Similar to the way an optical telescope increases our ability to see faint stars, an X-ray telescope can concentrate the radiation from an X-ray star onto an electronic eye. An imaging detector can observe several black holes emitting X-rays simultaneously, or take pictures of very far away regions of diffuse X-ray emission.

Some of the clusters of stars in the universe are spherical where thousands or millions of stars are closely bundled together in different regions of the universe. Stars are set apart from each other by a few lightyears at the center of a cluster and farther apart towards the outer regions. Some of the celestial objects in the cluster radiate X-rays. Are those sources of X-rays massive black holes?

Figure 6. An Illustration of Star Clusters.

At the center of galaxies there are massive star cluster encompassing millions of stars. Currently, it is theorized that there is a super massive black hole at the center of every galaxy. One super massive black hole has been identified at the center of our own Milky Way galaxy. Stars huddle toward the center of the Milky Way and spread out toward the outer regions. However, the closeness of the stars may allow for some collisions or cohesions. Black holes may also be agents behind some of these enormous collisions through their tremendous gravitational effects.

Recently, astronomers have taken the first-ever image of Sagittarius A*, the supermassive black hole at the center of the Milky Way, is our own dedicated supermassive black hole, the rotating hub or core around which our galaxy revolves. The images were taken by a global consortium of radio telescopes known as the Event Horizon Telescope. The idea was that the more matter a black hole accretes, the brighter is the accretion disk, so, the increased energy and momentum ejects gas outward. At the center of the Milky Way there are very energetic sources of microwaves, also theorized to be coming from the area of the supermassive black hole. Astronomers estimate that there could be more than a hundred million black holes roaming around the stars in our Milky Way galaxy, but they have not yet concretely identified or detected an isolated or rogue black hole. Black holes are minuscule compared with the galaxies where they are theorized to be found, reaching less than one percent of the mass of a typical galaxy and a volume a billion times smaller than the volume of the galaxy.

Black holes are theorized to accrete nearby stars and other celestial objects. Astronomers monitoring Sagittarius A* for many years have estimated the mass of the supermassive black hole to be about 4.31 ± 0.38 million solar masses. Sagittarius A* is about twenty six thousand six hundred and forty light-years away from Earth, which is fortunately very far away. The observation and measurement of Sagittarius A* may provide evidence to support the hypothesis that supermassive black holes grow by accreting nearby matter, mass, and energy.

There have been speculation that eventually after billions of year supermassive black holes will accrete all other celestial bodies and material. There is that possibility at a very slow rate of accretion, but it has not happen yet in the lifetime of our universe, since there are still billions and trillions of celestial bodies and galaxies. It has been speculated that black holes are the center of galaxies and the cores of clusters, instead of the consumers of everything.

Clusters are not so far away from Earth. The Hyades star cluster is the closest star cluster to Earth, located only one hundred and fifty light-years away. As of yet, black holes at the center of clusters are only speculated, technology keeps improving our ability of observing deeper into the center of clusters, to detect X-rays, or gravitational waves, in the future.

Black holes serve a constructive purpose in our universe that we are beginning to understand such as assembling stars and celestial bodies, clusters, and galaxies, in the great expanse that are crucial to lifeforms.

Where else are the black holes found? As far as detecting the presence of black holes, astronomers may have better luck with binary systems. It has been theorized that if the distance and period of rotation is known for a binary star system then its mass can be calculated. If a binary star is small but its calculated mass is large, it is probable that the companion star may be undergoing its collapsing process, which may be the process of a dwarf star. The massive star and its companion may end up collapsing into two binary black holes that accrete matter, mass, and energy, in the surrounding spatiotemporal medium, that may be observed.

It has been theorized and detected that binary black holes are often divided into stellar binary black holes, formed either as remnants of a high-mass binary star system, or by dynamic processes and mutual capture, or from binary supermassive black holes, believed to be a result of galactic mergers.

Black holes have been found that have merged, even though they originated at different locations in the universe, spin in opposite directions, producing gravitational waves that can be detected.

There are astrophysical celestial objects with physical attributes that emit X-rays. For example, an active galactic nucleus is a compact region at the center of a galaxy that has a much-higher-than-normal luminosity with attributes that indicate that the luminosity is not produced by stars.

The celestial objects that emit X-rays include: galaxy clusters, black holes in active galactic nuclei, supernova remnants, binary stars and massive stars, containing a white dwarf, neutron stars, or black holes from X-ray binaries. Some celestial objects like moons emit X-rays, even our Moon, although most of the X-ray of the Moon originate from the reflected brightness of X-rays from our Sun.

There also celestial objects in our solar system which have been detected as X-ray emitters due to either the reflection of solar X-rays from their surfaces, charge exchange of their atmospheres with the highly ionized solar wind, and solar-wind initiated auroral emission, include the planets Venus, Earth, Mars, Jupiter, Saturn, and Pluto, the Io Plasma Torus, a blue cloud that is a region of higher concentration of ions and electrons located at Io's orbit. and at least two of the moons of Jupiter, and some comets.

For binary stars, a double emission of an X-ray have been detected as the stars orbit their gravitational center. The X-rays from a binary star system come from the area around the collapsed star where the material that is falling into the collapsed star is heated to an extremely high temperature.

It may not be achievable to always identify the companion star of a binary star even if the star is a source of intense X-rays once it becomes a black hole. The companion star may not be observable, and it may be very small. If it is too massive and unobservable to be a neutron star, it may be deduced that it is a black hole. Red giants are dying stars in the final stages of stellar evolution. It is theorized that our sun will eventually turning into a red giant in about five billion years, expanding and engulfing the inner planets in the solar system.

A red giant may expand spilling its matter and mass into its black hole companion which would explicate how a black hole becomes an intense X-ray source. The distance between the companion star and its black hole of a binary system is crucial. The more distant the companion star from its black hole, the greater the masses of the stars need to be to have a short orbital period, to develop a black hole from the massive collapsing star. It has been hypothesize that the distance between the companion star and the black hole may be at least a few thousands light-years for black hole formation to occur.

Moreover, as one of the binary stars collapses into a black hole, its companion star would be battered by material and energy as it orbits through all the debris from the supernova explosion prior to the formation of the black hole, even though the companion may be at a great distance, even if it is a few light-years, from the black hole.

The temperature of the companion star may collect debris, increasing its temperature, as it orbit through the dusty spatiotemporal medium. The gravitational effect between the two celestial bodies may decrease from the enormous loss of mass after the immense supernova explosion of one of the binary stars. The end result would be an observable star that orbits a gravitational center opposite to a tremendously intense detectable source of X-rays.

It may be speculated that in some neutron star, the magnetic field is strong and capable to prevent the formation of an accretion disc. The temperature of the material in the disc increases to an extreme temperature due to friction, emitting X-rays. The angular momentum of the material in the disc gradually decreases and falls into the contracted star. Moreover, additional X-rays are generated in the neutron star when the material buffets their surfaces. The X-ray emission from black holes is not constant, it varies in luminosity at intervals. The change in luminosity may provide information about the size of the black hole. So, the presence of these X-rays indicate the presence of a black hole or a neutron star, the X-rays emission in intermittent pulses, microwave pulses, indicating a black hole source from a neutron star. The X-rays from a black hole vary intermittently as the matter or mass is accreted in extensive numbers or amounts, other times infrequently.

Rotating neutron stars are called pulsars. Pulsars appear to emit pulses of radiation at regular intervals if the beam of particles points toward the Earth; they have very strong magnetic fields that eject jets of particles out along its two magnetic poles. The accelerated particles are emitted and may be observed as very strong luminous beams.

The closest pulsar to Earth is identified as 4715, a millisecond pulsar, rotating at 173 times per second, as the pulsar spins very fast and very stable, like a reliable clock or cycle counter.

Skylab was a science and engineering laboratory that was launched into Earth orbit by a Saturn V rocket in 1973. Skylab conducted detailed X-ray studies of the Sun. A faint X-ray experiment "S150" and source survey was performed. The experiment was mounted atop the upper stage of the Saturn 1B rocket which orbited briefly behind and below Skylab. The entire upper stage went through a series of preprogrammed maneuvers, to scan firmament about four degrees per minute, to allow the scanner to sweep across selected regions of the firmament. The pointing direction was determined using a computer, using the inertial guidance system of the upper stage combined with data from two visible star sensors which formed part of the survey. There were galactic X-ray sources observed with the scanner. The experiment was designed to detect minute photons or faint luminosity.

By the 1980s, the study of X-ray sources continued to be conducted, data was obtained from satellites that operated until the 2000s, among them, but not limited to, EXOSAT, RXTE, Ginga, HEAD Program, ROSAT, and ASCA. These satellites detected the first afterglow of a gamma-ray burst, contributing further data to enhance our understanding of the nature of these X-ray sources and the mechanisms by which the X-rays and gamma rays are emitted.

These mechanisms can shed light on understanding the fundamental physics of our physical reality in our universe. Observing the celestial objects in our universe with X-ray and gamma-ray instruments, it is possible to gather crucial data to answer our questions such as how the evolution of our universe progressed, how the universe actually began in the first place, or where it is heading.

§ 2. Hawking's micro black holes.

Micro black holes, also called quantum mechanical black holes, are hypothetically very small black holes, for which quantum mechanical effects play a significant role. The concept of micro black holes may exist that are smaller than a stellar mass was introduced and promoted by the British physicist Stephen Hawking. The mass of a micro black holes ranges from the minuscule Planck mass to more than trillions of kilograms, but the micro black holes are still a lot smaller than massive stellar black holes.

A massive astrophysical black hole has been detected at the center of our galaxy. It is theorized that black holes, in general, are everywhere. They may be at the center of star clusters, as part of binary systems. There are vast numbers of astrophysical massive black holes and it is hypothesized that there is also a vast number of micro black holes throughout the universe, including within particles, the critical particle may orbit the black hole an arbitrarily large number of times with arbitrarily long proper time.

There may be billions of galaxies in the universe. They must exist in the trillions, evenly distributed everywhere there is matter or mass, or elsewhere in the spatiotemporal medium. So, black holes are a staple of the universe!

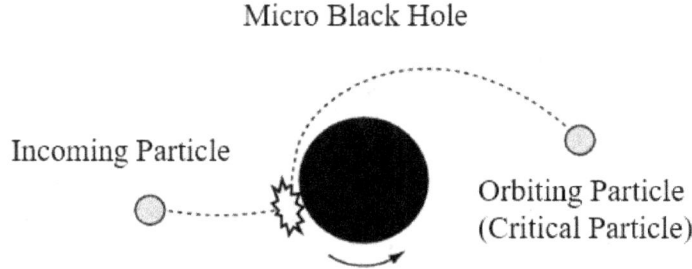

Figure 7. An Illustration of a Micro Black Hole as a Particle Accelerator.

It may be speculated that black holes naturally accelerate particles. Nevertheless, for a Schwarzschild-size micro black hole, which is a static spherically symmetric micro black hole, the center-of-mass energy of two particles of equal rest mass which have been at rest at infinity is not much more than its rest energy, which cannot be regarded as a high energy particle accelerator.

So far, our discussion has been focused on the supermassive black holes that are produced by very massive stars, which are the most observable black holes. The average stellar black hole has a mass between three to ten solar masses. Supermassive black holes exist in the center of most galaxies, including in our own Milky Way Galaxy. Supermassive black holes are enormously massive, with masses ranging from millions to billions of solar masses.

The evidence for the existence of supermassive black holes does not mean that black holes with lesser masses might not exist, because as it is usually theorized only stellar black holes may have the gravitational field strong enough to cause a tremendous collapse on a star, breaking the atomic structures that allow the formation of a black hole. Nevertheless, black hole may form at different sizes and masses according the General Theory of Relativity. The ability of an object of mass to become a black hole involves the contraction of the mass within the Schwarzschild radius. Any object of any size has the same characteristic when its gravitational field strengthens after contraction, its volume is reduced within the critical Schwarzschild volume, and the object of mass rapidly collapses when the escape velocity of its outer surface, and volume, increases over the effective velocity of light, collapsing into a singularity at the center of the volume of the object.

Any planet has the ability to be contracted to a volume smaller than the critical Schwarzschild volume to become a black hole. If the planet Venus in our solar system is contracted to a diameter less than a centimeter, or about the size of a hazelnut (a filbert). If we take Mount Aconcagua, the highest point in the Andes Mountains and in the Western Hemisphere at 22,831 feet (6959 meters), located in the Southern Andes Mountains, with its peak is in Argentina, and its western flanks in the coastal lowlands of Chile, just north of Santiago, Chile, and by some very advanced technological means contracted within a critical Schwarzschild volume to become a black hole, the singularity might be about the size of an atomic nucleus. The smaller the object of mass the smaller the singularity, until the size of the singularity reaches an atomic scale where there might be physical reason(s) for limitation on the mass density.

According the Stephen Hawking, the force of the Big Bang may be capable of contracting small objects of mass to micro black holes at the time that the universe was created. The force of the gravitational field of the small object of mass alone would not be enough to achieve the compactification needed for a micro black hole. During the Big bang, there was the possibility for mass and matter collisions with enormous forces and pressure that may have compacted and contracted the mass or matter into a volume smaller than a critical Schwarzschild volume to the point where gravitation is so intense that a black hole is formed. Furthermore, Hawking also proposed the idea that micro black holes evaporate rapidly before they can gain any significant mass.

2.1. Black hole collisions.

Black hole collisions may happen when the two or more black holes get close enough to one another to gravitationally influence each other. The spatiotemporal medium is enormous, so, the collisions are typically the products of a binary system consisting of two massive stars.

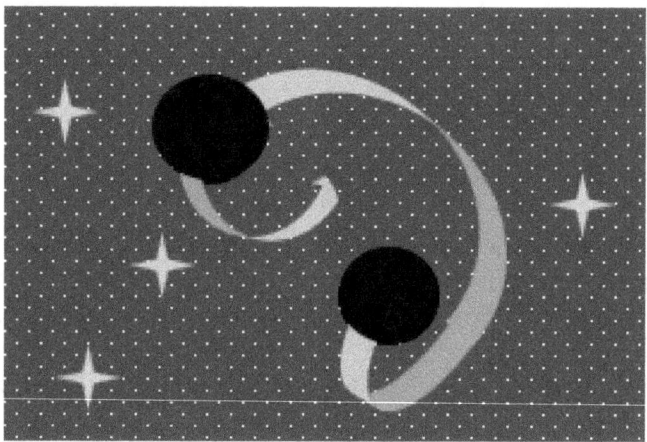

Figure 8. An Artistic Illustration of a Potential Collision of Two Black Holes.

It is possible for two or more black holes to collide if they come close enough to one another that they cannot escape the strong gravitational fields among them, they will eventually merge into a larger black hole. When two black holes or more spiral around one another and finally collide, the black holes emit gravitational waves, ripples in the fabric of space-time, these spatiotemporal ripples can be detected with extremely sensitive equipment. Black holes and black hole mergers are typically dark, which makes these events unobservable to telescopes used by astronomers. Nonetheless, it is theorized that a black hole merger may produce light when it influences nearby material to radiate.

2.2. Did astronomers recently find the fastest spinning black hole in our universe?

The case involves two black holes much larger than our sun, with the larger black hole having about forty solar masses. The binary pair was first observed in 2020 when LIGO detected a blast of gravitational waves from a large collision, denoted GW20019, between two black holes. The precise location of the two black holes is not exactly identified.

A group of researchers determined that the merger between the two black holes was asymmetrical and massive, with gravitational waves blasting out of the collision in one direction while the newly merged black hole was likely ejected out of its galaxy at more than 3 million miles per hour (4.8 million kilometers per hour) in the opposite direction. This observation proposes that the two black holes had a disorderly association before their rough merger. Two black hole objects pushed each other in a narrowing orbit, they began to sway, as they also precess several times per second. The bigger black hole was about forty times more massive that our sun, spinning as fast as it is physically achievable.

The effect of the precess was estimated to be ten billion times faster than any other precessing effect previously measured. The precessing effects were predicted by the General Theory of Relativity for the massive objects in our universe. So, this type of rough merger of two black holes may not be as unusual as previously thought. So, this black hole collision may have been extremely unusual, or the accepted models of black hole collisions need to be updated.

2.3. The spatiotemporal curvature versus the gravitational potential for celestial bodies.

The positions of various celestial bodies are shown in the following figure on a "curvature versus gravitational potential" graph representing the spatiotemporal locations of planets, moons, and black holes.

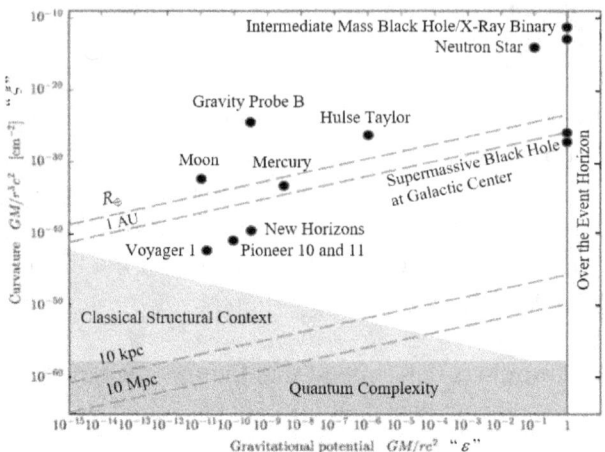

Figure 9. Curvature versus Gravitational Potential Graph.

The graph shows a volumetric parameter to quantify the strength of a gravitational field. Adapted from (Psaltis, 2008) Looking at the Schwarzschild space-time, it is natural to measure the strength of the gravitational field at a distance "r" away from an object of mass M by the volumetric parameter "ε".

$$\varepsilon = \frac{GM}{rc^2} \triangleq \frac{Spatiotemporal\ Volume}{Dimensional\ Volume} \quad (2.1)$$

$$\varepsilon \triangleq \frac{V_{Structural\ Context}}{V_{Quantum\ Complexity}} \equiv Volumetric\ Index \quad (2.2)$$

Which is proportional to the Newtonian gravitational potential and is also directly related to the redshift. *Quantum Complexity Volume is Information at the Planckian Scale, while Structural Context Volume is a Complex Language at the Classical Scale.* A Complex Language provides a framework for a gravitational system of classical complexity and principles for significance or algorithms in the physical processes of reality.

Quantum or Planckian gravitational fields correspond to the limit $\varepsilon \to 0$, which leads to Minkowski's nearly flat space-time of the Special Theory of Relativity. Weak gravitational fields correspond to

$\varepsilon \ll 1$, which leads to Newtonian gravity. Strong gravitational fields, that are accessible to an observer, are characterized by $\varepsilon \to 1$, at which point the black-hole horizon of an object of mass M is approached. Please note that formally the event horizon of a Schwarzschild black hole occurs at $\varepsilon = 2$; but the factor of 2, in this context, will be dismissed, in favor of the emphasis on "structural versus dimensional" arguments.

In the case of a strong gravitational field, the strong gravitation is characterized, not by a large gravitational potential, that is, a high value of the volumetric parameter "ε", but instead by a large spatiotemporal curvature given by

$$\xi \triangleq \frac{GM}{r^3 c^2} \triangleq \frac{1}{r^2} \qquad (2.3)$$

$$\xi \equiv Spatiotemporal\ Curvature$$

$$\varepsilon \equiv r^2 \cdot \xi \qquad (2.4)$$

$$r \equiv \sqrt{\frac{\varepsilon}{\xi}} \equiv \sqrt{\frac{Volumetric\ Index}{Spatiotemporal\ Curvature}} \qquad (2.5)$$

Because the condition(s) that the curvature needs to satisfy in order for a classical gravitational field to be considered "strong" depends on the particular deviation from the General Theory of Relativity under examination, the spatiotemporal curvature parameter "ξ" will not be normalized to any particular energy density value, but rather it will be left as a dimensional value from this point forward. The Ricci scalar is set as follows, $R \sim \xi$. If the distance "r" is larger than the radius of the object, then the Ricci scalar of the General Theory of Relativity vanishes.

The spatiotemporal curvature parameter is an appropriate parameter with which to measure the strength of a gravitational field in a geometric theory of gravity, such as the General Theory of Relativity, because the spatiotemporal curvature is the lowest order variable of the gravitational field that cannot be set to zero by a coordinate transformation. Moreover, because the spatiotemporal curvature measures energy density, a limit on spatiotemporal curvature will correspond to an energy scale beyond which additional gravitational degrees of freedom may become important.

On the previous graph, the horizontal axis indicates the gravitational potential. The vertical axis indicates the spatiotemporal curvature of the gravitational field at a radius "r" away from a central object of mass M.

Three-dimensional regions of the volumetric parameter "ε" may be referred to as a "parameter volume" with a potential greater than one representing distances from a gravitating object of mass that is less than the volume of the event horizon and is inaccessible to observers. The red vertical line on the right-hand side of the graph marks the limit of the event horizon. The above graph is intended to be a schematic, and is not meant to show an extensive number of objects and systems that have been used, or could be used, to assess the veracity and applicability of the General Theory of Relativity.

The region of the scale tests in the solar system is broadly bounded by the Moon, Gravity probe B, Mercury, and the Pioneer and New Horizons spacecrafts. The Voyager spacecraft has been included in the graph, even though it was never applicable for tests of the General Theory of Relativity. (Will, C.M., 2015) and (Everitt, C.W.F. et Al., 2015) The famous Hulse-Taylor binary pulsar, is a neutron star binary, it is also roughly in the region of tests for the solar system. (Hulse, R.A. et Al., 1975) Black holes and neutron stars are in the strong gravitational field regime, what is more, black holes are at the boundary limit of the event horizon. Adapted from (Psaltis, 2008).

The gravitational potential, that is accessible to observers, is bounded by the line $\varepsilon \simeq 1$. There is no limit to the maximum curvature that can be observed, except for the Planck limit, where $r \simeq 1.6 \times 10^{-33}$ *cm*. The dimensional length "r" is at many orders of magnitude above the boundaries of the above figure. The General Theory of Relativity is regarded by some as a complete theory, which can also be applied below the Planck scale. The test of the General Theory of Relativity in the trans Planckian medium has not been feasible yet.

The bottom regions on the graph represent trans Planckian or Planckian regions where "A Dynamic Theory of Space-Time" proposes that spatiotemporal wavelets interfere, also known as complexity, to allow the emergence of quantum gravitation from spatiotemporal geometry or entanglement. Larger spatiotemporal waves can also interfere at the classical level as validation of the General Theory of Relativity, where Pressure $\equiv$ Energy Density. (Nieves, 2020)

2.4. What is the minimum curvature in the spatiotemporal medium?

The curvature for the homogeneous and isotropic regions of the universe decreases as the universe expands. The undisturbed Friedmann–Lemaître–Robertson–Walker metric is applicable in isotropic and homogenous spatiotemporal regions. The present universe has a curvature which is just above the boundary of the region marked quantum complexity, since cosmological quantum complexity curvature is not yet completely dominant over underpressurized mass. Nonetheless, the curvature will approach this limit asymptotically. This concept represents a new paradigm, a fundamental minimum cosmological curvature scale, the scale is shown in the above graph by the region labeled "quantum complexity."

Galaxies are useful astrophysical instruments for the measurement of the General Theory of Relativity. Their innermost regions can be modeled as test particles orbiting a central supermassive black hole. Galactic rotation curves, which can only be explained by introducing classical spatiotemporal wave interference, describe galaxy velocities up to the outermost regions. Systems below a constant acceleration scale of approximately 1.2×10^{-10} m/s^2 could also be modeled adding a contribution to the gravitational filed in the form of classical spatiotemporal wave interference. The constant acceleration is the diagonal boundary of the region labeled "classical spatiotemporal wave interference" in the above graph.

Note that the regions of the parameter "ε" volume occupied by classical spatiotemporal wave interference and quantum complexity overlap, and there is enhancement between these two novel components of the cosmological model. However, it is possible to compare between their effects, since their properties are similar. In particular, classical spatiotemporal wave interference forms clumps just like baryonic matter, while quantum complexity does not.

The General Theory of Relativity has not been evaluated in the region between curvatures of $\sim 10^{-40}$ and $\sim 10^{-50}$. This spatiotemporal region corresponds to the region between a solar system scale and a cosmological scale probed by galaxy surveys and the Cosmic Microwave Background. It is very challenging to find systems which span these scales. In theory systems do exist, in the form of galaxies, clusters and superclusters. The galaxy rotation curves transition from Schwarzschild orbits in their innermost regions, to their outer regions dominated by classical spatiotemporal wave interference. However, observations are obstructed by the fact that the observations are limited by the resolution of current telescopes, so only the outer regions can be observed. Moreover, the region situated between a region where the General Theory of Relativity is extremely well-constrained by solar system tests and a region where classical spatiotemporal wave interference and quantum complexity have to be evaluated to be utilized, and where we must take the cosmological model into account, because the effect of a large-scale structure on a background of quantum complexity becomes complementary. It is theorized that this stacking effect may be at the origin of the observed cosmic acceleration.

The following cosmological tests of The General Theory of Relativity may be classified as:

1. The tests of consistency between the growth of large structures in the universe and the development of the universal expansion. A correspondence in the equation of the volumetric parameter "ε" of quantum complexity, inferred from the two approaches, can indicate a resurgence of the General Theory of Relativity based on the cosmological paradigm of an emergent and smooth quantum complexity.

2. The detailed measurements of the linear growth factor across different scales and redshifts. The comparison of the cosmological mass distribution, inferred from different probes, especially from redshift spatiotemporal distortions and lensing.

Chapter 7

The Invisible Dark Star.

Black holes are great conceptual labs to probe into and test our best theories or hypotheses about the nature of space-time. A Black hole is a staple of the Theory of General Relativity, providing an empirical instrument to analyze the classical and quantum attributes of our physical reality. It is very important to understand the basic concepts of black holes and the spatiotemporal sites and surrounding celestial bodies at their location in the universe.

§ 1. The enigma of a black hole.

A black hole has been called a riddle, inside of a puzzle, wrapped up by an enigma. It is one of the current enigmas of Astrophysics in our universe. It seems appropriate to start our journey with a captivating story that sparks the imaginations of children through science and learning about black holes with a clever science parody from science author for kids, Chris Ferrie, illustrated in a book by Susan Batori.

There was a black hole that swallowed the universe.
I do not know why it swallowed the universe – oh well, it could not get worse.
There was a black hole that swallowed a galaxy.
It left quite a cavity after swallowing that galaxy.
It swallowed the galaxies that filled the universe.
I do not know why it swallowed the universe – oh well, it could not get worse.

So far, scientists have struggled to know more about black holes since our civilization is getting started to explore the depths of the physical reality of our universe. There is fear about black holes and their destructive power, but also there is hope that black holes are much more helpful and creative as people learn more about the mysterious celestial objects.

Scientists are observing black holes with powerful telescopes that currently orbit the earth. Black holes have been identified by the gravitational effect they have on surrounding celestial bodies, and by the existence of gravitational waves from stellar black hole collisions. However, the physical processes inside a black hole are inaccessible to the traditional optical telescopes, but some technological advances are providing and alternate way to observe the radiation from a black hole so that astronomers and physicists can theorize what might be the processes that occur inside the event horizon of a stellar black hole. On the other hand, theoretical physicists are trying to probe deeper into the enigma of black holes with instruments like the General Theory of Relativity, Quantum Mechanics, and the latest astronomical technologies.

§ 2. The astrophysics of a massive black hole.

2.1. Not even light can escape.

A fundamental definition of a black hole is a spatiotemporal region where gravitation is so strong that nothing, including light or other electromagnetic waves, can escape it. The General Theory of Relativity predicts that a sufficiently contracted mass can warp space-time to form a black hole. For a Schwarzschild black hole, the radius is given by $R_s = GM/c^2$, any object of mass, matter or energy, that falls through the event horizon at the Schwarzschild radius, cannot escape the gravitational field of the black hole, not even light.

2.2. Is a dark star also a black hole?

The impressive idea of a "Dark Star" was conceived and publicized in 1783 by an English country parson named John Michell, according to recorded history. John Michell was a brilliant and imaginative natural philosopher, or scientist, even though his name remains forgotten by the passage of time, his significant research and contribution to physical science remains alive and well.

So, it is still possible to review, recognize, and appreciate this amazing man, this allegedly little, short man, of black complexion, unpretentious, and chubby. John Michell studied at Cambridge University, where he later taught Hebrew, Greek, Mathematics, and Geology. John Michell became rector of Thornhill, near Leeds, where he did most of his important research, and was visited by other scientists, including the well-known American scientist Benjamin Franklin.

Among his scientific achievements are: the description of how the magnetic force exerted by each pole of a magnet decreases with the square of the distance in 1750, authoring a book that helped establish seismology as a science in 1755 after a major earthquake in Lisbon, Portugal, suggested that earthquakes spread out as waves through the solid Earth and are related to the offsets in geological strata now called faults. After all these achievements he was elected to the Royal Society in 1760.

Nonetheless, it is his hypothesis about a "Dark Star" in 1783 that transposes John Michell, the scientist, forward to our future, when his incredible idea becomes an active field of study that involves breakthroughs in the understanding of physical reality, space, and time. His method, still the same, examines the mass of a star, that due to a strong gravitation, collapses into a Dark Star. In other words, the star becomes a supernova. The birth of a stellar black hole from gravitational collapse.

It is also worthy to mention that thirteen years later, another brilliant scientist and mathematician in France, Pierre-Simon Laplace, Marquis de Laplace, who in 1796 wrote in his Exposure of the System of the World: "A luminous star of the same density as the earth and whose diameter would be two hundred and fifty times greater than that of the sun, would not, by virtue of its attraction, allow any of its rays to reach us; it is therefore possible that the largest luminous bodies in the universe are, by that very fact, invisible". The birthdate of Michell's Dark Star had been corroborated by another great mind.

Furthermore, John Michell had the insight that Isaac Newton's conception of light consisted of minuscule particles of matter. Today, it is accepted that photons have relativistic mass, and just like Michell thought possible they emanate from a star, and their velocity may be decreased by the gravitational field of the star, as they travel to the Earth and other celestial bodies.

John Michell also had the insight that the star's mass may be calculated from the decrease of the velocity of the particles. He wondered about the intensity of the effect, and speculated about the critical escape velocity from the star.

So, finally, he had the insight: What if the Dark Star's gravitational field were so strong that the escape velocity was greater than the velocity of light? John Michell knew about Ole Roemer, a Danish astronomer, that calculated the speed of light to be about 220,000 kilometers/second, about 26% lower than the true value of 299,792 kilometers/second, by observing the eclipses of Jupiter's moon during the years 1668–1674. Consequently, John Michell realized that even light would have to fall back to the star's surface. The calculation of the escape velocity relied on the mass of the Dark Star, the escape would be greater than the velocity of light for a supermassive star that was five hundred times the solar mass, assuming that the star's average density was equal to the sun's.

In such a scenario, light could not escape the collapsing star, so the Dark Star would be invisible to an observer. A realization that was far ahead of its time!

However, they did not know at the time that light travels at a constant velocity, regardless of the strength of the gravitational field, due to the changes of scale of space and time. In other words, the observer may see no change of scale within the observer's universe due to the laws of quantum mechanics.

2.3. Are the laws of Quantum Mechanics contradictory?

So, what is an observer?

An observer does not have to be a human being that measures or witnesses an event, it only means that a reference has to be chosen when defining the values of a measurement.

The act of observation is simply the entanglement of two systems, where the complexity, or information, that describes each system is not independent. Consequently, there is one set of complexity, or information, describing both things as a single system.

The above definition of observation circumvents key problems in Quantum Mechanics: decoherence, the measurement problem, and Wigner's friend. The method of why the wave function collapses has not been explained adequately; for example, why do some interactions collapse and some interactions only decohere? How could two observers (Wigner and friend) disagree on the quantum state of a system?

If the observer is part of the system, any subsystem can be an observer. No collapse takes place but subsystems may become entangled. Anything in the multiverse is reversible.

Figure 1. The Everett View according to the First Law of Quantum Mechanics.

The observer is outside the system. Observations irreversibly collapse the wave function.

Figure 2. The Copenhagen View according to the Second Law of Quantum Mechanics.

Quantum Mechanics has two different laws to describe how a system changes as time passes. The first law acts most of the time, and describes how the spatiotemporal waves expand, or contract, as they flow smoothly through spacetime.

Rule I:

"Except during a measurement of position or momentum, the spatiotemporal wavelet expands or contracts smoothly, and deterministically."

Hence, this law endows the quantum system with the characteristic to explore simultaneously distinct possible outcomes and the alternate histories of alternate realities of alternate universes for all possible outcomes available to the smooth flow of the spatiotemporal wavelet. *The first law is an interdimensional law for all possible realities and outcomes of an arbitrary event in spacetime.*

The second law involves the distinctive circumstance of measurement. During a measurement, a microscopic system interacts with a quantum system, to allow a single outcome to be manifested, but disallowing the probabilities of distinct outcomes to take place in spacetime.

Rule II:

"During a measurement of position or momentum, the spatiotemporal wavelet collapses around the position, or momentum, where is being measured, with a probability that is approximately equal to the square of the height of the wavelet, or more accurately with a probability that is equal to the area of the hemisphere facing the measurement for a spherical spatiotemporal wavelet."

The second law involves probabilities which leads to contradiction with the first law under the present paradigm of Quantum Mechanics. This contradiction may be superseded by considering *the first law as interdimensional, or multiversal, a superposition of states for the distinct possible outcomes and alternate histories of alternate realities of alternate universes*. Each state with probability one, as the measuring device is able to observe distinct outcomes of each alternate history.

The second law treats the quantum system as being in a definite outcome, and the measuring device has observed only that outcome, with each distinct outcome having some definite probability.

Hence, *the second law is a universal law, not an interdimensional law, or not a multiversal law, for the interaction of a microscopic system with a quantum system, in reference to a single universe of the possible multiverse.* As in the law of Max Born.

Therefore, Quantum Mechanics has been very successful in its prediction and applications. By shifting the paradigm of the laws of Quantum Mechanics, the measurement problem may be solved. The outcome of that realization and its confirmation is that Quantum Mechanics will be compatible with realism.

2.4. Are the correlations of Quantum Mechanics due to spacetime?

If the spins of two entangled particles are measured in two orthogonal directions, the spins would have the same kind of reciprocal uncertainty than position and momentum. According to the uncertainty principle of Quantum Mechanics, if you measure the spin in one direction, you cannot assign an exact simultaneous value for the other spin. Hence, the uncertainty principle of Quantum Mechanics would assert a fundamental limit to the precision, through any of several mathematical inequalities, with which the values for the spins of the two entangled particles in two orthogonal directions can be predicted from initial conditions.

So, if one were measuring the spin of one of the particles in the up-down vertical direction, and the other particle in the left-right horizontal direction, there would not be correlation between the simultaneous measurements. However, if the measurements are performed in the same direction, then the measurements would be maximally correlated. If one of the two orthogonal directions of the measurements is adjusted toward a parallel direction, it is possible to determine how strongly correlated the measurement outcomes become if the spins are determined already before particle decay. In such case, the measured correlation would have the upper bound of Bell's inequality.

Quantum mechanical experiments have shown that such upper bound can be violated. A difficulty arises if the measurement value for the spin has been determined when the entangled state of the two particles was created, because one could not explain the observed correlations between the spins of the two particles. If the measurement value was not determined at entanglement, then the spins would become determined non-locally on both sides the moment one measures at least one of the spin values. Such a correlation is stronger than it could possibly be if the spin had been determined before the measurement. Thus, these types of measurement results oppose determinism and locality in favor of unpredictability and non-locality.

Let us imagine two entangled particles that travel away from each other, at the same speed of light, in a parallel direction through a spatiotemporal dimension in isotropic and homogeneous spacetime. The advanced temporal waves between them follow parallel paths from their instant localities to their origin of entanglement. Even though the spatiotemporal paths may not be exactly alike in terms of, but not limited to, expansion, contraction, or torsion, it is hypothesized that the advanced temporal waves, as well as its tachyons, traveling between the two entangled particles, would provide the information exchange or handshake effectively, when a measurement is performed on either particle, since the paths are directions of the same dimension.

If the paths of the particles were through two orthogonal dimensions, the advanced temporal waves, or tachyons, would not follow their directional paths through the same dimension. It is hypothesized that advanced temporal waves, or the phases of their tachyons, between two entangled particles, may not be equally affected by spatiotemporal perturbations, when traveling through the directions of different dimensions. As a result, a measurement on either particle would be uncertain and the correlation of measurements would be reciprocally uncertain. However, if one of the two orthogonal directions of the measurements is adjusted toward a parallel direction, as one of the directional paths becomes less orthogonal and more parallel to the other, the measurements may become correlated at a lower bound of Bell's inequality since the advanced temporal waves, or the phases of their tachyons, may be less affected by the spatiotemporal perturbations of the direction of the orthogonal dimension.

Additionally, if the spins had been determined at entanglement, it is hypothesized that the observed correlations between the spins of the two particles are due to the deterministic conditions of the particles that exists at the locality of origin which are strongly conserved by the advanced temporal waves and their tachyons through the directional paths of the particles. Causality would be preserved when measurement is performed before particle decay. Otherwise, the spins would become determined non-locally on both sides the moment one measures at least one of the spin values. Hence, if these hypotheses are supported, the correlations of Quantum Mechanics and Causality may be directly related to spatiotemporal wave theory.

2.5. How could black holes be described today?

The Schwarzschild radius represent the numerical value advanced by John Michell and Pierre Simon Laplace to formulate the critical radial distance of a collapsed star of mass M.

$$R_S = \frac{2GM}{c^2} \simeq 3\frac{M}{M_\odot} \quad (Kilometers) \qquad (2.1)$$

where "R_S" is the critical Schwarzschild radius, "$M_\odot$" is the solar mass. Any celestial body of mass M contracted within the Schwarzschild critical volume becomes a black hole.

The development of the wave theory of light obscured the hypotheses about the original dark star, and disincentivized the numerical analysis of the gravitational action on the propagation of light, discouraging the pursuit of those valuable forward-looking concepts. The General Theory of Relativity ascribed the attributes of light to gravitation and spatiotemporal curvature and geometry, inspiring new ideas and provoking deeper insight into black holes.

Let us use a tesseract of light to depict and relate its geometry, orientation and deformation to the spatiotemporal medium. A tesseract, also known as a hypercube, is a four-dimensional representation of a six-dimensional hypercube. It may be described as an extension of the idea of a spatiotemporal square to a six-dimensional space-time in the same way that a spatial cube is the extension of the idea of a square to a three-dimensional space. The tesseract can be unfolded into six diverging or converging temporal cubes from three spatial cubes, just as each spatial cube can be unfolded into six diverging or converging temporal cubes from a diverging cubic singularity of light that may beam or dim at the center, depending on the direction of the arrow of time. Let us imagine that the square from any of the sides of a tesseract is an infinitesimal square on a large spatiotemporal spherical wave that is nearly flat at the scale of the quantum square, as the wave diverges at the speed of light in all directions.

The representation of the hypercube of light represents the entire history of a nearly flat quantum surface on a diverging or converging wave, resulting from the interference of quantum spatiotemporal wavelets, in multidimensional diagram. If the outer temporal cube is removed, the remaining cube, a past cube, is spatial, the spatiotemporal cubic singularity, the emerging point, is the future of now. The expanding or converging cubes of light originate a tesseract or hypercube of light. All linear velocities of the hypercube of light are based on the speed of light which in turn defines the spatial unit of length and the unit of time of "c." Light travels on all linear paths at forty five degrees from all Cartesian coordinate axes of space or time.

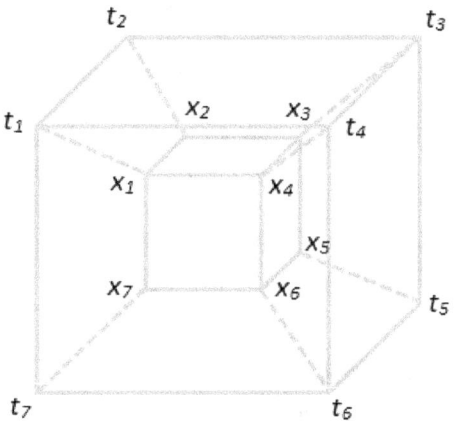

Figure 3. The Tesseract of Light.

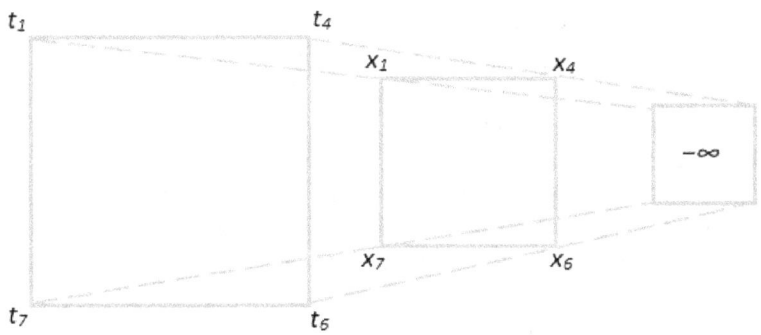

Figure 4. The Spatial Divergence of Each of the Six Sides of the Tesseract of Light.

The tesseract of light permits us to represent the spatiotemporal causality of the spatiotemporal volume of the medium. For example, in flat space-time the tesseract of light retains its orientation and undeformed shape, which is the background medium of the Special Theory of Relativity. At the event "now" of a cosmic tesseract, the light ray, between the past and future cosmic cubes (shaded), leads the way for a measuring quantum tesseract, on its own reference frame, that travels toward its future.

The light ray spans the spatiotemporal medium that connect each corner of the past and future cubes. All objects of mass, matter, or energy, traveling slower that the speed of light, are restrained to the spatiotemporal volume of the cosmic tesseract, regardless of their path through the spatiotemporal medium. The volumetric velocity of divergence for the hypercube is given by $dx^n/cdx = nx^{n-1}/c^{n-1}$, where "$n$" is the number of dimensions, and the volumetric acceleration is $d^2x^n/c^2dx^2 = 2nx^{n-2}/c^{n-2}$ in terms of the outer cube. Aside, if the Earth were cubic, the gravitational acceleration at the center of each face of the cubical Earth would be 0.006 ms^{-2} lower than the value at the surface of the spherical Earth.

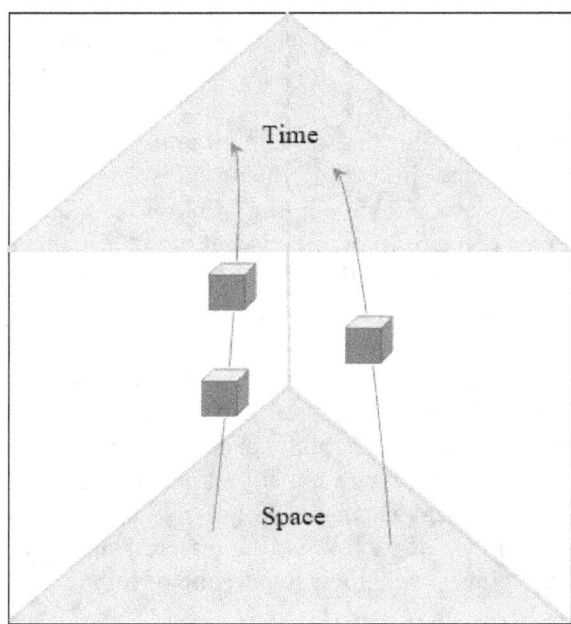

Figure 5. The Tesseracts Travel Through Minkowski's Space-Time.

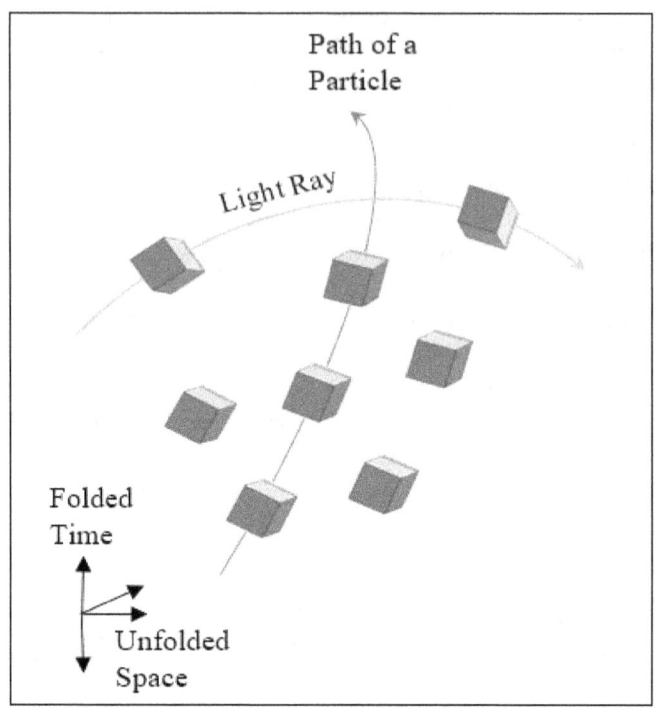

Figure 6. The Tesseracts Travel Through the Malleable and Curved Space-Time of GR.

The speed of light remains the same in all non-accelerating frames of reference in free space-time; consequently, tesseracts would have the same shape, as they would have in the Minkowski space of the Special Theory of Relativity, which is flat but still malleable. If there is gravitation present, the spatiotemporal medium gets curved as the spatiotemporal medium in the General Theory of Relativity.

Einstein's' equivalence principle is also applicable to a six-dimensional spatiotemporal medium, where the gravitational field of mass, matter, or energy, bends both space and time, and deforms tesseracts as they travel though the curvature of the medium. Nonetheless, the Special Theory of Relativity continues to be valid in the frame of reference of each traveling tesseract and any particle that may be traveling inside their hypercube of light even after deformation.

2.6. The collapsing dark star.

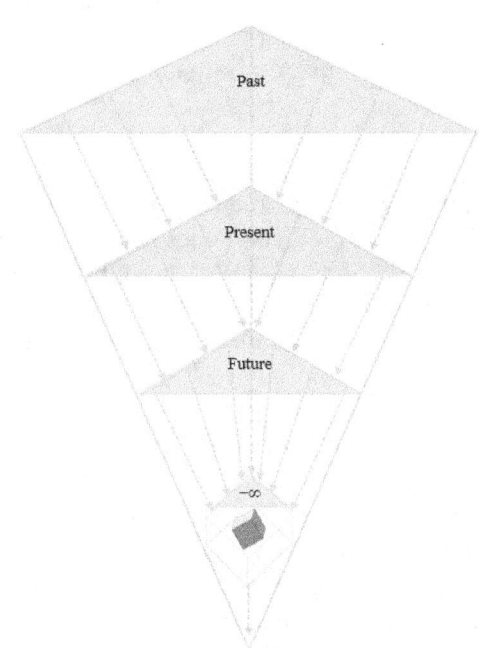

Figure 7. A spatiotemporal diagram illustrating a converging complex tesseract due to the gravitational collapse of a massive star into a black hole.

The figure above illustrates the classical structure of the spatiotemporal complexity of a collapsing star during the theoretical formation of a massive black hole. The figure illustrates a partial record of the collapse of a spatiotemporal complex tesseract from its beginning convergence to a series of stages toward a multidimensional singularity. The outer temporal side of the tesseract is shown collapsing inward, three temporal dimensions, each temporal dimensions is the conjugate of a spatial dimension. Space is on the vertical direction, measured downwards during convergence. Time is measured horizontally and vertically upwards during divergence, which would be the opposite case. The singularity will be at the center of the forming classical structure where the radius approaches infinity. The curvature of the spatiotemporal medium is represented by the orientation and torsion of the inner spatial cube and/or the temporal spatial cube of the spatiotemporal hypercube or complex tesseract, as the cubes follow the paths of light rays.

A black hole may form from an asymmetric gravitational collapse, while spatiotemporal deformations of the event horizon are rapidly disperses as gravitational radiation, or spatiotemporal waves. The event horizon oscillates according to the quasi-normal modes. Eventually, the black hole subsides into an axisymmetric equilibrium arrangement.

The spatiotemporal curvature of the hypercubes is nearly flat when the hypercubes are very distant from the reach of the gravitational field of the black hole; so, the hypercubes are not deformed and stay aligned to the path of the light rays. When the hypercubes are near the strong gravitational field of the black hole, the hypercubes are deformed and/or tilted inward by the spatiotemporal curvature and/or torsion.

At the Schwarzschild critical radius, $R_S = GM/c^2$, the hypercubes will tilt to forty five degrees and the resultant velocity vector points inward toward the center of the classical structure of the black hole; any mass, matter, and energy falling through the event horizon of the Schwarzschild critical volume would be oriented toward the singularity at the center. After the formation stage, the arrow of time and the spatiotemporal medium flip, space turns into time and time turns into space, as space converges toward the singularity.

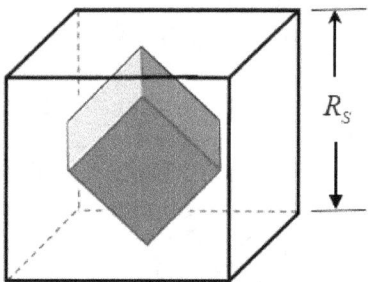

Figure 8. A Schwarzschild Hypercube.

The converging velocity and acceleration of a single hypercube in terms of the Schwarzschild cubic volume of the outer cube would be denoted as

$$\frac{1}{c^n}\frac{dR_S^n}{dR_S} \equiv -\frac{nR_S^{n-1}}{c^{n-1}} \qquad (2.2)$$

$$\frac{1}{c^n}\frac{d^2 R_S^n}{dR_S^2} \equiv -\frac{2nR_S^{n-2}}{c^{n-2}} \tag{2.3}$$

Substituting for the converging velocity and the acceleration, we obtain

$$v = \frac{3R_S^2}{c^2} = -\frac{3G^2 M^2}{c^6} \tag{2.4}$$

$$a = \frac{6R_S}{c} = -\frac{6GM}{c^3} \tag{2.5}$$

2.7. The formation of a theoretical white hole.

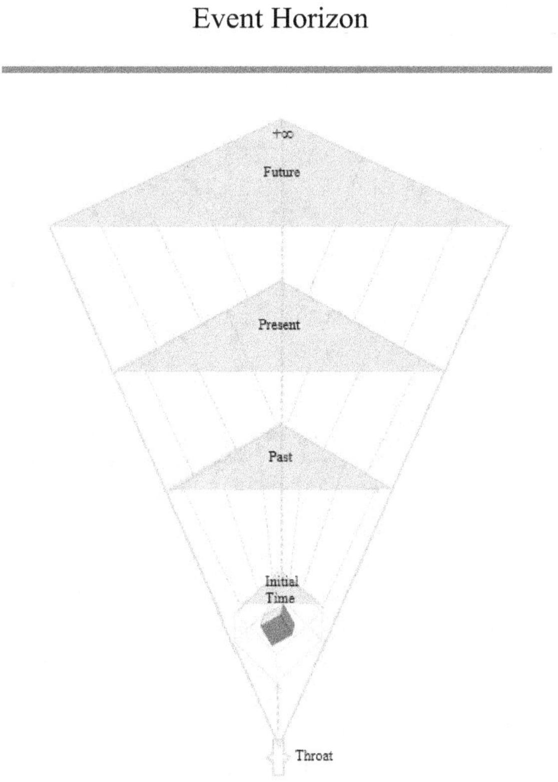

Figure 9. A spatiotemporal diagram illustrating a diverging complex tesseract due to the expansion of a theoretical white hole in a black hole - white hole pair.

§ 3. The tenses of time.

Time has a set of temporal forms taken by the spatiotemporal medium to indicate the duration, the continuance, or completeness of an action in relation to the initial instant. Therefore, time is complex and the conjugate of space. Each spatial dimension has a conjugate temporal dimension. (Nieves, 2020)

$$\textit{Apparent Time} \equiv \textit{Proper Time} \pm \textit{Time Dilation} \tag{3.1}$$

$$ict_A \equiv ic\tau_P - i\Delta c\tau_D \tag{3.2}$$

Inside the event horizon of the black hole, the radial coordinate "r" becomes temporal, and the temporal coordinates become spatial. For every particle that crosses the event horizon and inescapably falls toward the singularity at the center of the black hole in a radial free-fall along its trajectory as $r \to 0$, the proper time "τ" is measured by a comoving clock, minus the initial time "τ_0", is given by

$$\tau_P - \tau_0 = -\frac{4GM}{3c^2}\left(\frac{rc^2}{2GM}\right)^{\frac{3}{2}} \tag{3.3}$$

which behaves orderly at the event horizon. The apparent time, as measured by the clock of a distant observer, is given by

$$t_A = (\tau_P - \tau_0) - \frac{4GM}{c^2}\left(\frac{rc^2}{2GM}\right)^{\frac{1}{2}} + \frac{2GM}{c^2}\ln\left(\frac{\sqrt{\frac{rc^2}{2GM}}+1}{\sqrt{\frac{rc^2}{2GM}}-1}\right) \tag{3.4}$$

and diverges toward the singularity as $r \to \frac{2GM}{c^2}$.

The Schwarzschild coordinates, which only expands across:

$$\frac{2GM}{c^2} \leq r < +\infty \tag{3.5}$$

$$-\infty < t < +\infty \tag{3.6}$$

are not well suited for the analysis of the spatiotemporal causal structure.

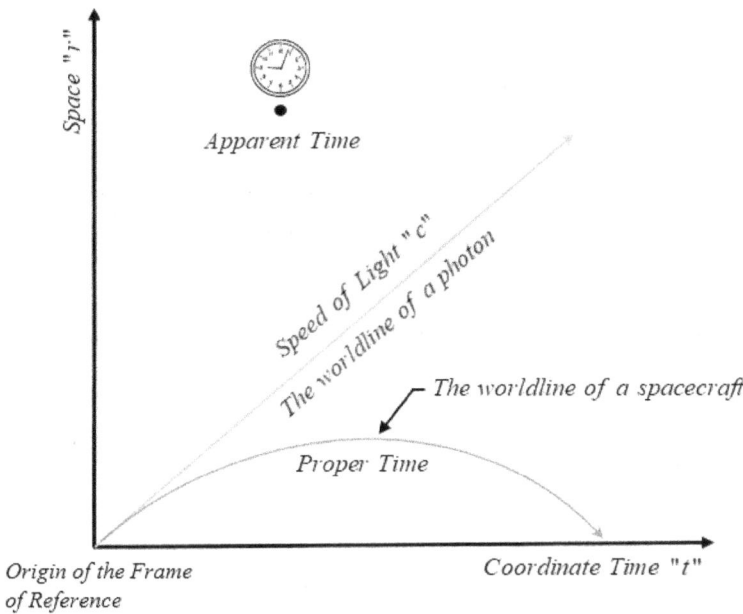

Figure 10. A space-time diagram illustrating proper time and coordinate time.

The above figure well illustrates the difference between the "proper time" for a spacecraft that travels from the origin frame of reference as measured by a clock placed on the surface of a planet, and the "apparent time," measured by an independent clock at a distant spatiotemporal point far from any gravitational field or spatiotemporal distortion. The proper time interval is the time it takes for the spacecraft to travel out to space, on a curved trajectory, to survey stars and return to the planet.

Let us express the Schwarzschild factor as

$$1 - \frac{R_S}{r} = 1 - \frac{2GM}{rc^2} \qquad (3.7)$$

All gravitational effects follow directly from the Schwarzschild equations. According to the General Theory of Relativity, the spatiotemporal medium around the geometry of a tesseract is described by the following Schwarzschild tesseract metric. Let us start with the Schwarzschild metric for Spherical symmetric body given by:

$$ds^2 = -\left(1-\frac{2GM}{rc^2}\right)dt^2 + \left(1-\frac{2GM}{rc^2}\right)^{-1}dr^2 + r^2 d\Omega^2 \qquad (3.8)$$

where $d\Omega^2 = d\theta^2 + \sin^2\theta d\varphi^2$ is the metric of a unit two-sphere. The solution describes the external gravitational field generated by any static spherical mass, whatever its radius. In the General Theory of Relativity, Birkhoff's theorem states that any spherically symmetric solution of the vacuum field equations must be static and asymptotically flat.

For Minkowski's space-time with unfolded time, or six-dimensional flat space-time, we have

$$ds^2 = -\left(1-\frac{2GM}{rc^2}\right)(dt_X^2 + dt_Y^2 + dt_Z^2) + \left(1-\frac{2GM}{rc^2}\right)^{-1}(dx^2 + dy^2 + dz^2) \qquad (3.9)$$

$$ds^2 = -\left(1-\frac{2GM}{rc^2}\right)dt^2 + \left(1-\frac{2GM}{rc^2}\right)^{-1}dr^2 \qquad (3.10)$$

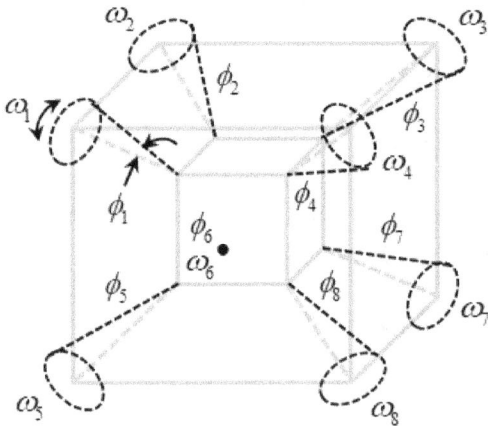

Figure 11. An illustration of the angles of curvature or torsion in a tesseract or hypercube.

Let us express the Schwarzschild tesseract or hypercube curvature factor as

$$1 - \frac{R_S}{r} \equiv 1 - \frac{2GM}{rc^2} \equiv 1 - \frac{\phi_S}{2\pi} \qquad (3.11)$$

Let us denote the Schwarzschild-Cartan tesseract or hypercube torsion factor as

$$\Omega_{SC} \equiv 1 - \frac{\omega_{SC}}{2\pi} \qquad (3.12)$$

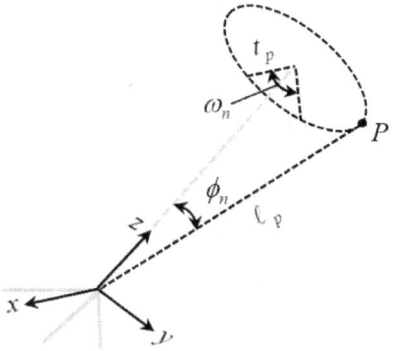

Figure 12. An illustration of the angles of the hypercube torsion and curvature factors, where $0 \le t_p \le t$ and $0° \le \omega_n \le 360°$. The slant height is given as the line element "$\ell_p \equiv ct_p$".

The sum of the angles of all the hypercube torsion factors "ω_{SC}" represents the average deformation due to the torsion on all eight diagonal rays passing through the corners of the hypercube. The initial angle of torsion is zero on a Schwarzschild undeformed hypercube.

$$\omega_{SC} \equiv \sum_{n=1}^{8} \frac{\omega_n}{8} \qquad (3.13)$$

The specific torsion factor on a single diagonal ray passing through one of the corners of the hypercube is given by

$$\omega_{SC} \equiv \omega_n \qquad (3.14)$$

The sum of the angles of all the hypercube curvature factors "ϕ_S" represents the average curvature on all eight diagonal rays passing through the corners of the hypercube. The initial angle of curvature is 0° on a Schwarzschild uncurved hypercube.

$$\phi_S = \sum_{n=1}^{8} \frac{\phi_n}{8} \tag{3.15}$$

The specific curvature factor on a single diagonal ray passing through one of the corners of the hypercube is given by

$$\phi_S \equiv \phi_n \tag{3.16}$$

Where the angle "ϕ_S" or "ω_{SC}" is measured in degrees or in radians. Aside, $1° \times \pi/180° = 0.01745$ rad, $\pi / \pi = 1$, and $360° = 2\pi$ rads. If the angle $\phi_S = 0°$, the spatiotemporal medium around the Schwarzschild hypercube is nearly flat similar to Minkowski's space-time. If the angle $\omega_{SC} = 0$ or nearly zero, the torsion is negligible. Also, if the sum of the angles inside the triangle is greater than or less than 180°, there is positive or negative curvature.

$$\phi_m \neq 0, \ \phi_n \neq 0, \ \phi_o \neq 0 \tag{3.17}$$

$$180° < (\phi_m + \phi_n + \phi_o) < 180° \quad \rightarrow \quad \textit{Spatiotemporal Curvature} \tag{3.18}$$

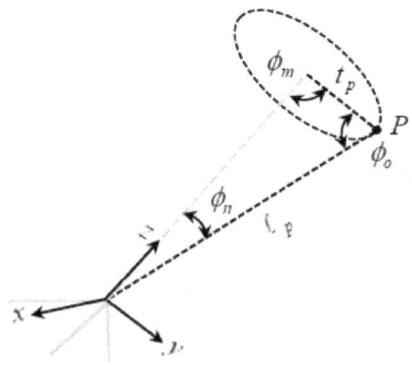

Figure 13. An illustration of the spatiotemporal curvature of the triangle of the hypercube or tesseract.

3.1. The curvature from a triangle.

It is interesting to point out that the sum of the angles inside a triangle on a Euclidean plane equals π. The total curvature of a geodesic triangle equals the deviation of the sum of its angles from π. The attributes of the Theory of the Special or the General Relativity spring from a Quantum Theory of Quantum Mechanics. The quantum relativistic attribute of the spatiotemporal wave function scales up to the classical world arounds us in our universe.

In its simplest definition, curvature is the numerical quantity by which a curve deviates from a straight line, or a curved surface deviates from a Euclidean plane. The curvature of a space, or for a surface, which is locally isotropic and homogeneous is described by a single Gaussian curvature. Curvature may be defined as the numerical quantity by which a Gaussian spatiotemporal surface deviates from being a Euclidean spatiotemporal plane. A Gaussian curvature can be defined without reference to an embedding space, such an intrinsically curved two-dimensional spatial surface is a simple example of a Riemannian manifold.

The Ricci scalar, or the scalar curvature, is the simplest curvature invariant of a Riemannian manifold. The scalar curvature is given by a single real number for each point of a Riemannian manifold determined by the intrinsic geometry near the point. *The scalar curvature may be described by the amount of change in the volume of a small geodesic ball in a Riemannian manifold from the same ball in a Euclidean space.*

Let us imagine a variable triangle in a metric space as a set of points in conjunction with a metric on the set. The function of the metric defines a distance between any two points of the set. The metric satisfies the topological properties of the spatiotemporal Gaussian surface. The curvature of the Gaussian spatiotemporal surface is an intrinsic measure that only depends on distances that are measured on the surface itself.

The variable triangular surface moves in such a way that the sum of the reciprocals of its intercepts on the three Cartesian coordinate axes is a constant *"d."* The interior angle of any one of its three corners is equal to the arc sine of two times the area of the triangle divided by the product of the lengths of the segments of the two sides of the corner that has the interior angle.

$$\theta_x = Arc\operatorname{Sin}\left(\frac{2 \cdot Area}{\overline{ac} \cdot \overline{ab}}\right) \quad (3.19)$$

Where a, b, and c are the intercepts on the x, y, and z axes, respectively.

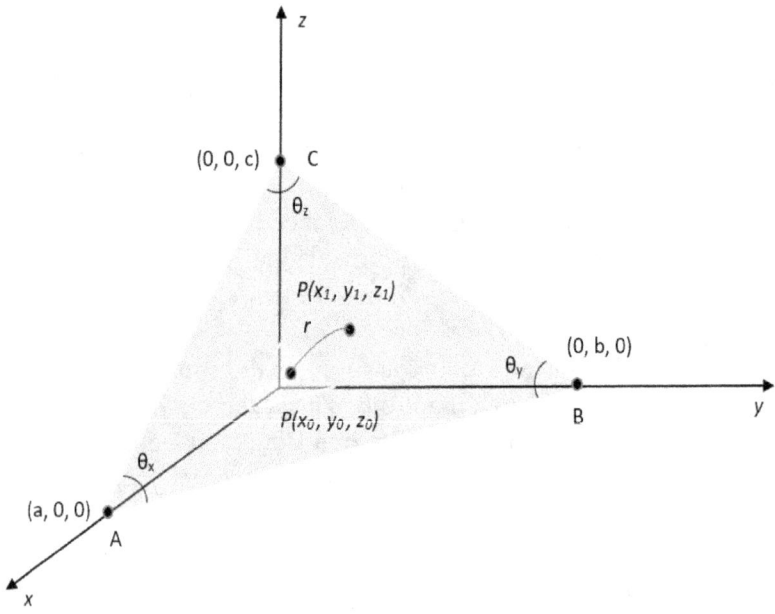

Figure 14. A variable triangular surface

Since the sum of the reciprocals of the intercepts on the Cartesian coordinate axes is equal to a constant "d," we have

$$\frac{1}{a} + \frac{1}{b} + \frac{1}{c} = d \quad (3.20)$$

where "d" is a constant whose equation can be denoted as

$$\frac{1}{a}\left(\frac{1}{d}\right)+\frac{1}{b}\left(\frac{1}{d}\right)+\frac{1}{c}\left(\frac{1}{d}\right)=1 \tag{3.21}$$

The above equation indicates that the triangular surface passes through the fixed point $P\left(\frac{1}{d},\frac{1}{d},\frac{1}{d}\right)$.

The triangular surface meets the Cartesian coordinate axes at A, B, and C, respectively, so that the intercepts on the x, y, and z axes are: $a = OA$, $b = OB$, and $c = OC$. Let us denote the triangular surface as $\vec{r}\cdot\vec{p}=d$. The spatial distance vectors at points A, B, and C, are: $a\vec{e}_x$, $b\vec{e}_y$, $c\vec{e}_z$.

Since the points A, B, and C lie on the triangular surface, we obtain

$$\vec{e}_x \cdot \vec{p} = \frac{d}{a} \tag{3.22}$$

$$\vec{e}_y \cdot \vec{p} = \frac{d}{b} \tag{3.23}$$

$$\vec{e}_z \cdot \vec{p} = \frac{d}{c} \tag{3.24}$$

Thus, let us subtitute for $\vec{r}$ to arrive at the intercept form of the variable triangular surface,

$$\vec{r} = x\vec{e}_x + y\vec{e}_y + z\vec{e}_z \tag{3.25}$$

$$\vec{r}\cdot\vec{p} = x\vec{e}_x \cdot \vec{p} + y\vec{e}_y \cdot \vec{p} + z\vec{e}_z \cdot \vec{p} = d \tag{3.26}$$

$$x\left(\frac{d}{a}\right)+y\left(\frac{d}{b}\right)+z\left(\frac{d}{c}\right)=d \tag{3.27}$$

Consequently, the intercept form of the variable triangular surface equation is given by

$$\frac{x}{a}+\frac{y}{b}+\frac{z}{c}=1 \tag{3.28}$$

It is interesting to note that the altitudes of an isosceles variable triangular surface intercept at the orthocenter. The altitude, perpendicular bisector, angle bisector, and median from the vertex angle to the base of an isosceles triangle are all the same segments.

The total curvature of a geodesic triangle on a Gaussian surface equals the deviation of the sum of its interior angles from π. If a surface has positive total curvature, the sum of the angles of a geodesic triangle on that surface will exceed π. If it has negative total curvature, the sum of the interior angles will be less than π. If a triangle is on a Euclidean plane, which is a flat surface with zero total curvature, the interior angles will sum to precisely π radians.

For higher-dimensional manifolds and surfaces that are embedded in a Euclidean space, the concept of curvature is complex, and it depends on the chosen direction for the manifold or surface.

The concept of radial scalar curvature relies on the ability to compare a curved space with another space that has zero curvature, or a constant curvature like a sector on a sphere.

Let us define the radial scalar curvature operator as:

The spatial radial curvature operator,

$$\Gamma_\nabla^2 = \pi^2 \cdot \nabla^2 = \pi^2 \left(\frac{\partial^2}{\partial x_0^2} + \frac{\partial^2}{\partial y_0^2} + \frac{\partial^2}{\partial z_0^2} \right) \tag{3.29}$$

The temporal radial curvature operator,

$$\Gamma_\odot^2 = \frac{\pi^2}{c^2} \cdot \odot^2 = \frac{\pi^2}{c^2} \left(\frac{\partial^2}{\partial t_{x_0}^2} + \frac{\partial^2}{\partial t_{y_0}^2} + \frac{\partial^2}{\partial t_{z_0}^2} \right) \tag{3.30}$$

The spatiotemporal radial curvature operator,

$$\Gamma_{SP}^2 = \Gamma_\triangledown^2 + \Gamma_\odot^2 \qquad (3.31)$$

Applying the operator on the complex wave function,

$$\Gamma_{SP}^2 \Psi(r,t) = \Gamma_\triangledown^2 \Psi(r) + \Gamma_\odot^2 \Psi(t) \qquad (3.32)$$

Let us now describe mathematically how it is possible to use a triangle as the radial measure of scalar curvature at an arbitrary quantum point or at a classical point from an origin of zero curvature with interior angles inside a triangle equal to θ_0 in a region of space-time.

Let us imagine a triangle on a spatiotemporal manifold that has no curvature with each corner on each of the positive axes x, y, and z. Then, let us also imagine a second triangle farther away on a spatiotemporal manifold with curvature, with each corner on each of the positive axes x, y, and z. The distance "r" of a proper spatial segment from the orthocenter point of the first triangle $P(x_0, y_0, z_0)$ to the orthocenter point of the second triangle $P(x_1, y_1, z_1)$ represents a radial distance of scalar curvature. Each interior angle of the triangle would be identified by the closest axis to it.

$$\vec{r} = |r| \angle \phi = |r| e^{\sqrt{\pi}(\phi)} \vec{a}_r \qquad (3.33)$$

For the positive axes chosen, we have

$$r = (r_1 - r_0) = \sqrt[2]{(x_1 - x_0)^2 + (y_1 - y_0)^2 + (z_1 - z_0)^2} \qquad (3.34)$$

$$\tau = (t_1 - t_0) = \sqrt[2]{(t_{x_1} - t_{x_0})^2 + (t_{y_1} - t_{y_0})^2 + (t_{z_1} - t_{z_0})^2} \qquad (3.35)$$

$$\phi = (\theta_1 - \theta_0) + (\theta_2 - \theta_0) + (\theta_3 - \theta_0) = \theta_x + \theta_y + \theta_z \qquad (3.36)$$

Curvature can be defined by the square of the derivative of the angle θ of a sector with respect to the length of its arc S.

$$S = r\theta \quad \text{(A Sector)} \qquad (3.37)$$

$$\frac{1}{r^2} = \left(\frac{\partial \theta}{\partial S}\right)^2 \equiv \left(\frac{\text{change in angle } \theta}{\text{change in radians}}\right)^2 \tag{3.38}$$

Therefore, we may describe curvature in a spatiotemporal wave as the square of the ratio of the change in the temporal coordinate distance with the change in the radian trajectory distance, to yield the reciprocal of the area of the surface of curvature. It is interesting to observe that the surface of curvature emerges as a temporal surface to manifest its reciprocal as a spatial curvature.

For the radial scalar curvature of a triangle when the change in radians equals π,

$$R = \frac{\partial^2 \left(-[\Psi(r)]^2\right)}{\partial r^2} = \frac{\partial^2 \left\{\left(\frac{1}{\ln e^{-r}}\right)\right\}}{\partial r^2} = \frac{\partial^2 \left[-\frac{1}{r}\right]}{\partial r^2} = \frac{1}{r^2} \tag{3.39}$$

$$R = \frac{\partial^2 \left(-[\Psi(r)]^2\right)}{\partial r^2} = \left(\frac{\partial(\theta_1 - \theta_0)}{\partial S}\right)^2 + \left(\frac{\partial(\theta_2 - \theta_0)}{\partial S}\right)^2 + \left(\frac{\partial(\theta_3 - \theta_0)}{\partial S}\right)^2 = \frac{\partial \theta_x^2}{\pi^2} + \frac{\partial \theta_y^2}{\pi^2} + \frac{\partial \theta_z^2}{\pi^2} \tag{3.40}$$

$$\pi^2 \left(\frac{\partial^2 \Psi_x^2}{\partial x_0^2} + \frac{\partial^2 \Psi_y^2}{\partial y_0^2} + \frac{\partial^2 \Psi_z^2}{\partial z_0^2}\right) = \partial \theta_x^2 + \partial \theta_y^2 + \partial \theta_z^2 \tag{3.41}$$

$$R = \frac{\partial^2 \left(-[\Psi(ct)]^2\right)}{c^2 \partial t^2} = \frac{\partial \theta_{t_{x0}}^2}{\pi^2} + \frac{\partial \theta_{t_{y0}}^2}{\pi^2} + \frac{\partial \theta_{t_{z0}}^2}{\pi^2} \tag{3.42}$$

$$\pi^2 \left(\frac{\partial^2 \Psi_{t_x}^2}{\partial t_{x0}^2} + \frac{\partial^2 \Psi_{t_y}^2}{\partial t_{y0}^2} + \frac{\partial^2 \Psi_{t_z}^2}{\partial t_{z0}^2}\right) = \partial \theta_{t_{x0}}^2 + \partial \theta_{t_{y0}}^2 + \partial \theta_{t_{z0}}^2 \tag{3.43}$$

Therefore, the radial curvature is reciprocal to the sum of the squares of the changes in the interior angles of the triangle. Radial scalar curvature increases if the sum of the changes of the interior angles is positive, positive curvature increases. Conversely, radial scalar curvature decreases if the sum of the changes of the interior angles is negative, negative curvature increases, as the interior angles are compared to the interior angles of the triangle at zero curvature in flat space.

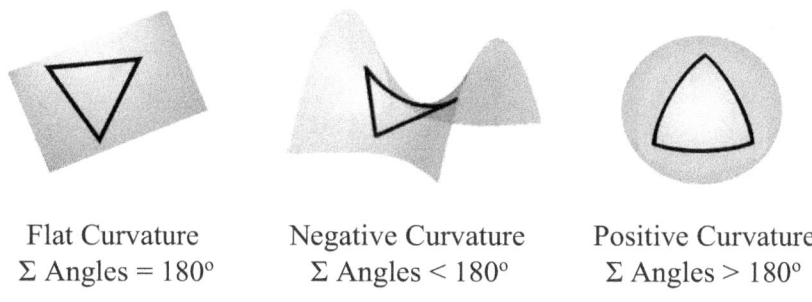

Flat Curvature   Negative Curvature   Positive Curvature
Σ Angles = 180°   Σ Angles < 180°   Σ Angles > 180°

Figure 15. An Illustration of Types of Curvature.

Where R is a real number representing the trace of the radial scalar curvature at a point $p(x, y, z)$ on the surface of a triangle, "r" is the radial distance, "τ" is proper time, π is a transcendental number, $\odot^2$ is the double tempus operator, $\Psi(r)$ is the spatial wave function equal to $1/\sqrt{r}$, but the second derivative of the negative or temporal squared wave function with respect to "r," is the radial scalar curvature of space, $d^2\left(-[\Psi(r)]^2\right)/dr^2 = 1/r^2$, and θ_n is an interior angle of a triangle that may or may not be geodesic.

Hence, *the wave function represents the curvature of space-time as a resultant wave produced by the interference of the spatiotemporal wavelets at every spatiotemporal point.* The more energy (fields or particles) that is present in the single wave function, the greater the frequency of the wave function, and the greater the curvature. In the total quantum system of multiple particles, even though each particle has its own wave function, there is a single total wave function representing all the particles.

3.2. The curvature from a hypercube.

Let us replace the coefficients of the temporal and spatial terms with the coefficient(s) of a Schwarzschild hypercube in degrees or radians in the line element.

The Schwarzschild metric for the hypercube curvature is given by

$$ds^2 = -\left(1° - \frac{\phi_S}{360°}\right)dt^2 + \left(1° - \frac{\phi_S}{360°}\right)^{-1} dr^2 \qquad (3.44)$$

$$ds^2 = -\left(\frac{\pi}{180°} - \frac{\phi_S}{2\pi}\right)dt^2 + \left(\frac{\pi}{180°} - \frac{\phi_S}{2\pi}\right)^{-1} dr^2 \qquad (3.45)$$

$$ds^2 = -\left(1 - \frac{\phi_S}{2\pi}\right)dt^2 + \left(1 - \frac{\phi_S}{2\pi}\right)^{-1} dr^2 \qquad (3.46)$$

The Schwarzschild-Cartan metric for the hypercube curvature and torsion is given by

$$ds^2 = -\left(1 - \frac{\phi_S}{2\pi} - \frac{\omega_{SC}}{2\pi} + \frac{\phi_S \omega_{SC}}{4\pi^2}\right)dt^2 + \left(1 - \frac{\phi_S}{2\pi} - \frac{\omega_{SC}}{2\pi} + \frac{\phi_S \omega_{SC}}{4\pi^2}\right)^{-1} dr^2 \qquad (3.47)$$

A spatiotemporal measuring device, a spatiotemporal bridge curvature-torsion detector or deformation detector, may be sent on a probe through a theoretically traversable black hole - white hole pair to record the curvature and torsion on the trajectory of the probe. Such a device may be constructed in the form of a complex tesseract, with a past hypocube, a symmetrical cube, and a future hypercube.

3.3. The Tolman-Oppenheimer-Volkoff counter-gravitational equation.

The Tolman-Oppenheimer-Volkoff equation named after the authors of the famous papers, governs the hydrostatic equilibrium of a perfect fluid as modeled by the General Theory of Relativity. Therefore, it can be applied to stellar objects that are static to predict certain limits on the masses and the sizes of stars. (Oppenheimer et Alia, 1939, and Tolman, 1934 and 1939)

The pressure of a static spherical object, like a star, which has the "Schwarzschild metric" in the outside spatiotemporal medium, satisfies the following differential equation called the Tolman-Oppenheimer-Volkoff equation.

$$\frac{dP}{dr} = -\frac{Gm}{r^2}\rho\left(1+\frac{P}{\rho c^2}\right)\left(1+\frac{4\pi r^3 P}{mc^2}\right)\left(1-\frac{2Gm}{rc^2}\right)^{-1} \quad (3.48)$$

Where "r" is a radial coordinate, and "ρ" and "$P(r)$" are the density and pressure, respectively, of the material at radius "r."

The equation is derived by solving the Einstein equations for a general time-invariant, spherically symmetric metric. For a solution to the Tolman–Oppenheimer–Volkoff equation, the Schwarzschild metric will take the form:

$$ds^2 = \left(1-\frac{2Gm}{rc^2}\right)c^2 dt^2 - \left(1-\frac{2Gm}{rc^2}\right)^{-1} dr^2 - r^2\left(d\theta^2 + \sin^2\theta d\phi^2\right) \quad (3.49)$$

Where "ρ" the mass density is a function of "r," and $m(r) = 4\pi \int_0^r r^2 \rho dr$. If the "$1/c^2$" terms are neglected, the Tolman–Oppenheimer–Volkoff equation becomes the Newtonian hydrostatic equation, used to find the equilibrium structure of a spherically symmetric body of isotropic material when general-relativistic corrections are negligible.

It is important to note that "ordinary differential equation" solving is not the most interesting thing in this process. An ordinary differential equation is an equation that involves some ordinary derivatives, as opposed to partial derivatives of a function, a differential equation whose unknown consists of one function of one variable and involves the derivatives of those functions. It is far more fascinating to know how one obtains the differential equation to begin with since it is not trivial. There are five methods for solving an ordinary differential equation: solution by inspection, variable separable, homogeneous, linear differential equation, and the general solution.

If the boundary is "r" at "R," continuity of the metric and the definition of $m(r)$ require that the density "ρ_0" is at the radius "R" of the star, and "M" is the total mass $M = m(R) = 4\pi \int_0^R r^2 \rho dr$.

Furthermore, if the mass is computed by integrating the density of the object over its volume, it will yield the larger value.

$$M = \int_0^R \frac{4\pi r^2 \rho}{\sqrt{1 - \frac{2Gm}{rc^2}}} dr \quad (3.50)$$

$$\frac{dm}{dr} = 4\pi r^2 \rho \quad (3.51)$$

$$m = \int_0^R 4\pi r^2 \rho \, dr \quad (3.52)$$

The difference between these two quantities will be the "gravitational binding energy" of the object divided by "c^2" and it is negative.

$$\delta M = \int_0^R 4\pi r^2 \rho \left(1 - \frac{1}{\sqrt{1 - \frac{2Gm}{rc^2}}} \right) dr \quad (3.53)$$

Let us assume a static, spherically symmetric perfect fluid to derive the Tolman-Oppenheimer-Volkoff equation from the General Theory of Relativity. The metric components are similar to those for the Schwarzschild metric.

$$\frac{dP}{dr} = -\frac{1}{r}\left(\frac{\rho c^2 + P}{2}\right)\left(\frac{2Gm}{c^2 r} + \frac{8\pi G}{c^4} r^2 P\right)\left(1 - \frac{2Gm}{c^2 r}\right)^{-1} = -\frac{Ee^{r_s}}{De^{-r_s}} \quad (3.54)$$

Where "D" is the converging diameter, and "E" is the diverging energy of the spherically symmetric perfect fluid. Extracting a factor of "G/r" and rearranging factors of 2 and "c^2" results in the Tolman–Oppenheimer–Volkoff equation:

$$\frac{dP}{dr} = -\frac{G}{r^2}\left(\rho + \frac{P}{c^2}\right)\left(m + \frac{4\pi r^3 P}{c^2}\right)\left(1 - \frac{2Gm}{rc^2}\right)^{-1} \quad (3.55)$$

Why is the Tolman-Oppenheimer-Volkoff equation counter-gravitational?

For most of its life, a star is in a static situation where protons in the core are fused to a bound state. *The bound state has a lower energy than the two separated particles, so highly energetic photons are emitted.*

The specific type of fusion that occurs inside of a star is known as proton-proton fusion. Inside a star, this process begins with protons, which are simply a lone hydrogen nucleus, and through a series of steps of the fusion process, these protons fuse together and are turned into helium.

It is important to note that these photons generate an outward pressure, a radiation pressure. This outward force per unit area plays a large role in counteracting the inward gravitational force. What increases photon energy? The photon energy increases and is directly proportional to its frequency and amplitude, and inversely proportional to its wavelength. Thus, as frequency and amplitude increase, with a corresponding decrease in wavelength, the photon energy increases. This effect is why active stars have basically no upper limit for the mass.

The intensity of high energy photons "I_γ" can be defined as the energy associated with photons emitted from a unit surface area in unit time ($J/m^2/sec$). As a beam of high-energy photons with incident intensity I_0 passes through matter, the loss of intensity may be calculated using the Beer-Lambert equation:

$$I_\gamma = I_0 e^{-\left(\frac{\mu}{\rho}\right)x\rho} \tag{3.56}$$

$$e^{-\left(\frac{\mu}{\rho}\right)x\rho} = \frac{I_\gamma}{I_0} \tag{3.57}$$

where "x" denotes the path length (cm), "μ/ρ" is the total mass attenuation coefficient (cm^2/g) and "ρ" is the density of the matter (g/cm^3).

In the above equation, the total mass attenuation coefficient considers the contributions from all three photon absorption processes: the photoelectric effect, the Compton scattering and the pair formation. Calculations of the total mass absorption coefficients for a wide range of photon energies absorbed by specific atoms, compounds and mixtures are available from the National Institute of Standards and Technology.

For a beam of high energy photons:

$$Pressure \equiv \frac{Intensity}{c} \equiv Energy\ Density \qquad (3.58)$$

$$P_\gamma = \frac{I_\gamma}{c} = \rho_\gamma c^2 \qquad (3.59)$$

Gamma rays fired at helium ions breaks the deuterium bound state emitting highly energetic photons that produce an outward gravitational pressure or counter gravitation. The proton-helium bound state has a lower energy than the separated proton and neutron which have a higher energy; so, *highly energetic photons are emitted, and exert a negative gravitational impulse orthogonal to their direction of travel, if the highly energetic photons travel through the elliptical path of a waveguide or a magnetic storage ring with beamlines, the negative gravitational impulse may be the basis for a useful tractor beam technology to tow a spacecraft or clean up debris in orbit around a planet. The outward momentum of photons, or light, can produce counter-gravitation.*

The counter-gravitational pressure would have to be directed upward, upward or tilted, from the ground on Earth, or other celestial body, to lift a spacecraft from the ground, to counteract the Earth's gravitational field, which is a downward pressure. However, jets, helicopters, aircraft with ducted fans, airships, hovercraft, or other forms of propulsion, may be used for travel in the atmosphere, or to ascend near space. In space, rocket planes, ion propelled spacecrafts, or other forms of space propulsion systems may be used. The flight deck of an aircraft carrier is the surface from which its aircraft or space craft would take off and land, essentially a miniature double sided airfield in the atmosphere, at sea, or in space. The underside, opposite to the antigravitational engines, on the top side of a traditional ship, may be used to position jets, rocket planes, or other flying vehicles, and hold flying vehicles upside down by design, due to the weak but attractive gravitational field, which may be adjusted accordingly. This type of propulsion system design may allow for flying counter-gravitational all-purpose carriers; the carriers could have a standard carrier underside, and a "double-sided flight" deck, like a flight-deck tunnel, in the middle for multi-purpose deployment, to the air, sea, or space. This technology and its related concepts may begin a new era of travel for the armed forces of the free world, commercial shipping, and tourism.

The strong nuclear force, one of the fundamental forces, is the force between two protons. The amount of strong nuclear force acting between two protons is the same as the amount of strong nuclear force acting between two neutrons. Protons are bound together in the nucleus of an atom as a result of the strong nuclear force. Neutrons are a type of subatomic particle with no charge; neutrons are neutral. Neutrons are bound like protons into the nucleus of an atom as a result of the strong nuclear force. Aside, there are no proton-proton or neutron-neutron bound states. The negative scattering length for a di-proton or a di-neutron would show that the two protons or two neutrons do not form a stable bound state. Moreover, because nucleons have spin of one-half, the wave function must be antisymmetric with respect to the interchange of the nucleons.

A deuteron is a bound state of a proton and a neutron with a binding energy of $B = 2.224 \pm 0.002\ MeV$, a very weakly bound system. A gamma ray "γ" of energy "E" may be aimed at a deuteron nucleus to break the bound state into a proton and a nucleus so the particles move opposite to the incident gamma ray. However, this opposite effect of the nucleons would not happen if the bound state equals the energy "E" of the incident gamma ray. Let us find what would happen if the energy of the gamma ray is greater than the energy of the bound state.

By the law of conservation of energy, the energy of the incident gamma ray minus the energy of the bound state equals the sum of the kinetic energies of a neutron and a proton in a deuteron,

$$Gamma\ Ray\ Energy - Bound\ State\ Energy \equiv \frac{p_{proton}^2}{2m_{proton}} + \frac{p_{neutron}^2}{2m_{neutron}} \quad (3.60)$$

$$\therefore m_{neutron} = m_{proton} = m \quad (3.61)$$

By the law of conservation of momentum, we have

$$p_{proton} + p_{neutron} = \frac{E}{c} \quad (3.62)$$

$$P_{neutron} = \frac{E}{c} - P_{proton} \tag{3.63}$$

If the energy "E" is equal to the energy of the bound state,

$$\frac{P_{proton}^2}{2m_{proton}} + \frac{P_{neutron}^2}{2m_{neutron}} = 0 \tag{3.64}$$

$$\therefore P_{proton} = P_{neutron} = 0 \tag{3.65}$$

Let us aim the gamma ray and beam the energy at the bound state,

$$E = \gamma_i + B_{state} \tag{3.66}$$

If it is assumed that the energy of the incident gamma ray "γ_i" plus the energy of the bound state is equal to "E," we find

$$\gamma_i = \frac{P_{proton}^2}{2m_{proton}} + \frac{P_{neutron}^2}{2m_{neutron}} = \frac{1}{2m}\left(P_{proton}^2 + \left(P_{proton} - \frac{E}{c}\right)^2\right) \tag{3.67}$$

$$2P_{proton}^2 - 2\frac{E}{c}P_{proton} + \left(\frac{E^2}{c^2} - 2m\gamma_i\right) = 0 \tag{3.68}$$

Let us solve the above equation using the quadratic equation,

$$P_{proton} = \frac{-\left(-2\frac{E}{c}\right) \pm \sqrt{4\frac{E^2}{c^2} - (2)4\left(\frac{E^2}{c^2} - 2m\gamma_i\right)}}{4} \tag{3.69}$$

$$P_{proton} = \frac{2\frac{E}{c} \pm \sqrt{4\frac{E^2}{c^2} - 8\left(\frac{E^2}{c^2} - 2m\gamma_i\right)}}{4} \tag{3.70}$$

The momentum of the proton is a real value; therefore,

$$\sqrt{4\frac{E^2}{c^2} - 8\left(\frac{E^2}{c^2} - 2m\gamma_i\right)} \geq 0 \qquad (3.71)$$

$$\sqrt{4\frac{E^2}{c^2} - 8\frac{E^2}{c^2} - 16m\gamma_i} = \sqrt{-4\frac{E^2}{c^2} - 16m\gamma_i} \geq 0 \qquad (3.72)$$

Finding the minimum value if the expression is equal to zero, we have

$$4\frac{E^2}{c^2} = 16m\gamma_i \qquad (3.73)$$

$$\gamma_i = \frac{4}{16}\frac{E^2}{mc^2} = \frac{E^2}{4mc^2} \qquad (3.74)$$

Therefore, if the energy "$E = 2c\sqrt{m\gamma_i}$" is greater that the energy of the bound state, the gamma ray may produce, from the bound state, separate neutrons and protons that may emit highly energetic photons, depending on the amplitude and the frequency of the incident gamma rays.

$$E > B_{state} \qquad (3.75)$$

These highly energetic photons generate an outward pressure, a radiation pressure per unit area, producing a curvature of an outward secondary counter-gravitational field that is opposite to the inward primary gravitation field.

The counter-gravitational field equations of the highly energetic photons are given by

$$\gamma_{\mu\nu} - \left(\frac{1}{n-1}\right)\gamma c_{\mu\nu} = \frac{8\pi q_p}{\hbar} L_{\mu\nu} \qquad (3.76)$$

$$\Gamma_{\mu\nu} = \frac{8\pi q_p}{\hbar} L_{\mu\nu} \qquad (3.77)$$

Where "$\gamma_{\mu\nu}$" is the photon curvature tensor, "γ" is the trace of the photon curvature tensor, "$c_{\mu\nu}$" is the photon metric tensor, "n" is the number of spatiotemporal dimensions, "q_p" is the charge per particle of the particles in the bound state whose energy is released for counter-gravitation, where "$q_p \equiv \ell_p \cdot t_p$", "$L_{\mu\nu}$" is the stress-momentum-energy photon tensor, and "$\Gamma_{\mu\nu}$" is the photon counter-gravitational tensor. (Nieves, 2020)

$$\Gamma_{\mu\nu} = \gamma_{\mu\nu} - \left(\frac{1}{n-1}\right)\gamma c_{\mu\nu} \tag{3.78}$$

In the presence of a primary gravitational field, we have

$$\Gamma_{\mu\nu} - G_{\mu\nu} = \frac{8\pi q_p}{\hbar} L_{\mu\nu} \tag{3.79}$$

$$\Gamma_{\mu\nu} - G_{\mu\nu} - \Lambda g_{\mu\nu} = \frac{8\pi q_p}{\hbar} L_{\mu\nu} \tag{3.80}$$

Where "$G_{\mu\nu}$" is the Einstein tensor of the primary gravitational field, and "Λ" is the cosmological constant.

Applying values of μ/ρ for water taken from the above figure of 0.06323 cm^2/g at 1.25 MeV, as an approximate value for the range from 1.1732 to 1.3325 MeV gamma rays emitted by ^{60}Co, illustrates the effect of the depth of water ($\rho = 1$ g/cm^3) on the intensity (I) of these gamma rays. (Biswas, 2016)

It is important to note that high-energy photons lose their intensity exponentially, and unlike high-energy electrons, do not have a finite range as they pass through matter.

The following graph shows the photon energy of gamma rays (MeV) and the total mass attenuation coefficient (μ/ρ)

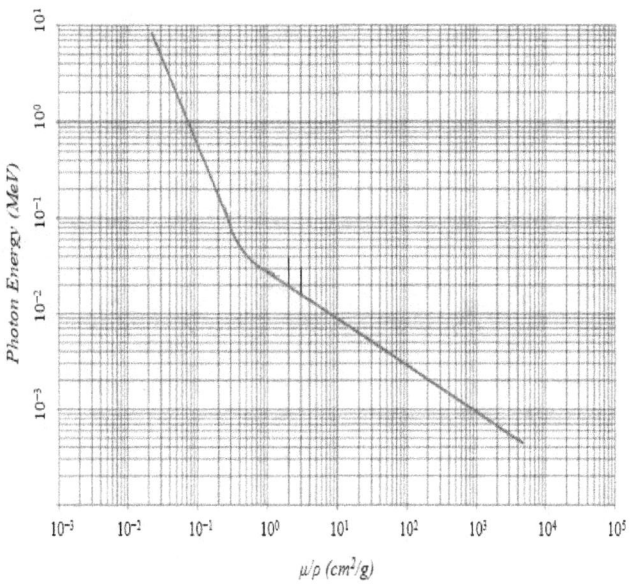

Figure 16. Photon Energy versus Mass Attenuation. (Biswas, 2016)

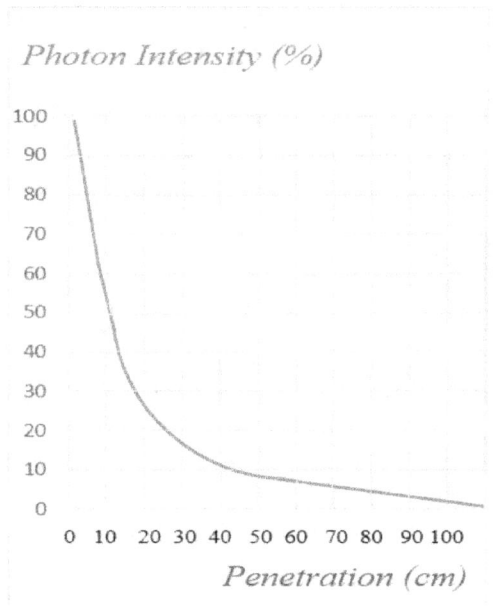

Figure 17. Penetration of High-Energy Photons in Water. Adapted from (Woodhead, 2013).

Conclusively, the higher the photon intensity, the lesser the penetration of water and the more is the pressure of the momentum on the water per square area, depending on the frequency and amplitude of the photon source.

Aside, a recent collaboration of Chinese and Japanese astrophysicists has reported the highest energy photons ever seen: gamma rays with energies up to 450 trillion electron volts (TeV). Aside, energy of a gamma ray photon: $> 2 \times 10^{-14}$ *J*, wavelength: $< 1 \times 10^{-11}$ *m*, frequency: $> 3 \times 10^{19}$ *Hz*. Synchrotron light is ideal for researching methods to take advantage of the high intensity, tunable wavelength, collimation, and polarization of synchrotron radiation.

Gamma rays are the most energetic radiation on the electromagnetic spectrum. Gamma rays are released in nuclear reactions and particle collisions. Gamma rays formed in this manner are about an order of magnitude less energetic than their cosmic-ray originators, which means that those cosmic rays reached energies far in excess of one peta electronvolt (10^{15} *eV*). Young supernova remnants are bright sources of energetic photons and neutrinos through the collision of particles accelerated inside the remnant. Interactions of accelerated particles in the expanding envelope, or in ambient radiation fields, will also produce secondary photons and neutrinos at some level.

From astronomical observations it appears that there are several possibilities for the end of the star. This is when the efficient fusion processes can no longer be maintained. If the original star is lighter than roughly 8 solar masses, the star will collapse to a white dwarf. The star does not produce energy anymore and cools down slowly. *The electron degeneracy at the core counteracts the gravitational force by means of the Pauli exclusion principle.*

If the white dwarf grows heavier than $1.44 M_\odot$, if the progenitor mass is larger than $8 M_\odot$, or if a white dwarf gathers mass from a neighboring star, the inward gravitational force is larger than the force outwards resulting from the electron degeneracy. *This means the static situation can no longer be maintained and as a consequence, the electrons in the core fuse with the protons and become neutrons. This releases a catastrophic amount of gravitational potential energy so generally a large part of the star's outer layers are blown away and a static neutron star is born. This explosion is called a supernova.*

Class	Energy (TeV)	Energy (eV)	Energy (µJ)	Frequency (yottahertz)	Wavelength (attometers)	Comparison	Properties
Photon	10^{-13}	1	1.602×10^{-13}	2.418×10^{-12} YHz	1.2398×10^{12} am	Near infrared photon	(for comparison)
Very High Gamma Rays	0.1	1×10^{11}	0.01602	24.2 YHz	12 am	Z boson	produces Cherenkov light
	1	1×10^{12}	0.1602	242 YHz	1.2 am	Flying mosquito	
	10	1×10^{13}	1.602	2.42×10^{3} YHz	0.12 am		Air shower reaches ground
	100	1×10^{14}	16.02	2.42×10^{4} YHz	0.012 am	Ping pong ball falling off a bat	Causes nitrogen to fluoresce
Ultra-High Gamma Rays	1000	1×10^{15}	160.2	2.42×10^{5} YHz	1.2×10^{-3} am		
	10 000	1×10^{16}	1602	2.42×10^{6} YHz	1.2×10^{-4} am	Potential energy of a golf ball on a tee	
	100 000	1×10^{17}	1.602×10^{4}	2.42×10^{7} YHz	1.2×10^{-5} am		
	1 000 000	1×10^{18}	1.602×10^{5}	2.42×10^{8} YHz	1.2×10^{-6} am		
	10 000 000	1×10^{19}	1.602×10^{6}	2.42×10^{9} YHz	1.2×10^{-7} am	Air rifle shot	
Planck Energy	1.22091×10^{16}	1.22091×10^{28}	1.95611×10^{9} J	1.885×10^{12} YHz	1.61623×10^{-17} am	Explosion of a car tank full of gasoline	

Table 1. Gamma Rays Characteristics. (Wikipedia, 2022)

The astonishing sudden opportunity for astronomers is photons. Every radioactive decay, from ^{56}Co to ^{56}Fe to ^{56}Ni, produces a gamma ray photon, the most energetic type of radiation.

However, the supernova is initially so dense that the gamma rays are contained. Before photons can escape, they travel around inside the rapidly expanding star and are converted to light that can be observed.

When each photon breaks free of its plasma containment, sometime after it was created, it may be the start of a very long journey.

Photons may travel through the universal space for billions of years before being absorbed by the eye of an observer.

Each photon carries one bit of hidden information: its wavelength, which determines its color and energy.

Photons that have energies greater than the band gap are absorbed, but the energy greater than the band gap is lost as heat.

3.4. Solving the Tolman-Oppenheimer-Volkoff equation for pressure or energy density.

Stars are held together by gravitation attraction exerted on each part of the star by all other parts, while the collapse of a star is resisted by internal thermal pressure.

These two forces play the principal role in determining stellar structure, they must be, at least almost, in balance.

All stars seem to be spherical and symmetric about their cores.

Stars may be isolated, static, and spherically symmetric. There are physical approaches to describe the structure of stars.

All the following physical approaches depend on the distance from the center of the star alone: the equation of hydrostatic equilibrium for each radius of a star which apply forces due to pressure differences that counter balance gravitation, the conservation of mass, the conservation of energy that applies to each radius, so that the change in the energy flux equals local rate of energy release, the equation of energy transport which is the relation between the energy flux and the local gradient of temperature, and the Tolman-Oppenheimer-Volkoff equation with an equation of state.

An equation of state $\rho = \rho(P)$ is used to solve the Tolman-Oppenheimer-Volkoff equation. It may be assumed that a contracted and fine-grained equation of state $\rho = \rho_0 =$ constant, which is a highly energetic case for the mass "M". With "ρ" as a constant, $m(r)$ can be explicitly found and simplification follows.

It is important to note that $dm \cdot dr = 4\pi r^2 \rho$, so, "$\rho$" can be replaced with the variables: P, dP/dr, m, dm/dr, which would create a complex equation.

It is possible to use $m = 4\pi r^3 \rho_0/3$, with a constant density, where there is no analytic solution for the general equation of state. Some factors of "r" are canceled and variables can be separated. Then, we get $dP/\sqrt{P} = f(r)dr$, and both integrals may be solved with straightforward techniques.

The mass inside a sphere is given by $m = (4\pi/3) \rho_0 r^3$ following the above assumptions, with the total mass related to the total radius in the same way: $M = (4\pi/3) \rho_0 R^3$. Then the differential equation can be rewritten as

$$\frac{dP}{dr} = -(P+\rho_0)\frac{\left(\frac{4\pi}{3}\right)\rho_0 r^3 + 4\pi r^3 P}{r\left(r - 2\left(\frac{4\pi}{3}\right)\rho_0 r^3\right)} = -\frac{4\pi}{3}(P+\rho_0)\frac{(\rho_0+3P)r}{1-\frac{8\pi\rho_0 r^2}{3}} \quad (3.81)$$

We now proceed with separation of variables. The boundary condition is: $P = 0$ at $r = R$. Hence, we can write,

$$\int_P^0 \frac{dP}{(P+\rho_0)(3P-\rho_0)} = \int_r^R \frac{rdr}{2\rho_0 r^2 - \frac{3}{4\pi}} \quad (3.82)$$

Both of these integrals are solvable using techniques from calculus,

The result of integrating is

$$\frac{1}{2\rho_0}\log\left(\frac{P+\rho_0}{3P+\rho_0}\right) = \frac{1}{4\rho_0}\log\left(\frac{\frac{8\pi\rho_0 R^2}{3}-1}{\frac{8\pi\rho_0 r^2}{3}-1}\right) \quad (3.83)$$

Using the earlier definition of "M" this is clearly equivalent to

$$\left(\frac{P+\rho_0}{3P+\rho_0}\right) = \left(\frac{1-\frac{2M}{R}}{1-\frac{2Mr^2}{R^3}}\right)^{\frac{1}{2}} \quad (3.84)$$

which can be solved for "P" to yield the desired result.

Aside, one could say $P = P_c$ at $r = 0$ and integrate in the opposite direction, finding a formula for pressure at any radius in terms of the central pressure and mass. The solution can then be inverted to find the radius in terms of the pressure, and its value at $P = 0$ would give the size of the object. This procedure is more sensible, if one has a better handle on the science responsible for the pressure, rather than on the actual sizes of objects, as it is often the case. For example, how much pressure is required from the counter-gravitational propulsion system of an aircraft or spacecraft to reach low Earth orbit?

For the pressure integral, one could complete the square to show that the integrand looks like the derivative of the inverse hyperbolic tangent, or equivalently the inverse tangent with $\sqrt{-1}$ in the appropriate places. Thus, it may be helpful to simplify things to use the inverse "1/tanh" instead of the tanh.

Nonetheless, if the inside of a star needs to be described, the line element, or metric, is utilized, not Newtonian physics. Even though it may not be advisable to assume that $m(r)$ is the enclosed mass, it would be interesting to demonstrate that it actually is the case. In this manner this amount of mass is related to astronomical observations of stellar objects, and its boundary conditions may be used. To demonstrate this, it is prudent to consider the total energy of a star that consists of: rest mass, internal energy, and potential energy. The mass density "ρ" is divided up in the following way,

$$\frac{p}{c^2} = \alpha_0 n + \left(\frac{p}{c^2} - \alpha_0 n\right) \tag{3.85}$$

in which the first term is the total rest mass that consists of "n" entangled particles per unit of volume with a mass of "α_0," and the second term consists of the internal mass density. Let us look into what all this represents for the potential energy.

According to the General Theory of Relativity, the proper volume of a Schwarzschild spherical shell of thickness "dr_S" is given by

$$dV_S = 4\pi r_S^2 \sqrt{|g_{rr}|} dr_S = 4\pi r_S^2 \left(1 - \frac{2m}{r_S}\right)^{-1/2} dr_S \tag{3.86}$$

The total rest mass "m_0" and the internal energy "U_S" inside a Schwarzschild sphere of radius "r_S" are denoted as

$$m_0 = \int_0^{r_S} \alpha_0 n \, dV_S = \int_0^{r_S} \alpha_0 n 4\pi r_S^2 \left(1 - \frac{2m}{r_S}\right)^{-1/2} dr_S \tag{3.87}$$

$$U_S = \int_0^{r_S} \left(p - \alpha_0 n c^2\right) dV_S = \int_0^{r_S} \left(p - \alpha_0 n c^2\right) 4\pi r_S^2 \left(1 - \frac{2m}{r_S}\right)^{-1/2} dr_S \tag{3.88}$$

If two mass amounts are subtracted from "m," the result would be the potential energy.

It is important to note that "$m(r)$" is the mass calculated according to The General Theory of Relativity.

$$U_S = \int_0^{r_S} 4\pi r_S^2 p \left(1 - \left(1 - \frac{2m}{r_S}\right)^{-1/2}\right) dr_S \tag{3.89}$$

$$U_S \approx \int_0^{r_S} 4\pi r_S^2 \left(\frac{r_S}{2\rho m}\right) dr_S \approx \int_0^{r_S} \frac{2\pi r_S^3}{\rho m} dr_S \tag{3.90}$$

The Newtonian limit is the last approximation, it seems to be equivalent to the gravitational potential energy. This finding suggests $m(r)$ may be the value of the total mass or energy, enclosed in a sphere of radius "r_S." This realization will be useful for a representation of a collapsing star, before saturation, to derive some limits on the size and the enclosed mass of a star collapsing toward a Schwarzschild singularity. The Tolman-Oppenheimer-Volkoff equation may have also been derived by observing the space-space component "rr" of the Einstein's Field Equations, where $\varepsilon = GM/r_S c^2$ signifies the volumetric index of the volumetric acceleration between the Newtonian limit and the Schwarzschild strong gravitational field regime "$\ddot{a}_N / \ddot{a}_S$", with respect to the previously defined variables:

$$G^r{}_r = r_S^{-2}\left(-1 + e^{-2\varepsilon} + 2r_S e^{-2\varepsilon}\phi\right) = 8\pi P \quad (3.91)$$

Consequently, the above conclusions render the equation for the gradient of the Tolman-Oppenheimer-Volkoff equation, including the previous definition of the enclosed mass:

$$\frac{d\phi}{dr_S} = \frac{m(r_S) + 4\pi r_S^3 P(r_S)}{r_S(r_S - 2m(r_S))} \quad (3.92)$$

Considering ϕ: $m/r_S = 1 - e^{-2\varepsilon} \ll 1$ in the Newtonian limit, and disregarding the contribution of pressure, this term reduces to $d\phi/dr_S = m/r_S^2$ which is the classical mass spatial acceleration. After integrating the conservation of the energy-momentum tensor, $\nabla_\mu T^{\mu\nu} = 0$, due of the Bianchi identity for the Einstein tensor, and the equation for the pressure "P," we obtain

$$-\frac{dP}{dr_S} = \left(\rho c^2 + P\right)\frac{d\phi}{dr_S} \quad (3.93)$$

Differentiating again with respect to "r_S", to obtain the counter gravitational field equation, we have

$$\text{Counter Gravitation} \equiv -\frac{\text{Spatiotemporal Pressure Acceleration}}{\text{Mass Density}} \equiv -\text{Temporal Curvature}$$

$$\frac{d^2\phi}{dr_S^2} = -\left(\frac{1}{\rho c^2 + P}\right)\frac{d^2 P}{dr_S^2} \qquad (3.94)$$

An elliptically polarized and accelerating gamma ray such that the tip of the electric field vector describes an ellipse in any fixed plane intersecting, and orthogonal to, the direction of propagation. An elliptically polarized and accelerating gamma ray may be resolved into two linearly polarized and accelerating gamma rays in phase quadrature, with their polarization planes at right angles to each other, since the electric field can rotate clockwise or counterclockwise as it propagates, elliptically polarized gamma rays display chirality. The accelerating spin operator of a gamma ray may be written as $\pm i(\partial^2/\partial\theta^2)$, and the spin angular momentum operator may be denoted as $\pm i\hbar(\partial^2/\partial\theta^2)$.

Circular polarization and linear polarization can be considered to be special cases of elliptical polarization. Substituting the equation for the gravitational field "ϕ" into the previous pressure equation, to produce the well-known Tolman-Oppenheimer-Volkoff equation:

$$\frac{dP}{dr_S} = -\frac{(\rho + P)(m(r_S) + 4\pi r_S^3 P)}{r_S(r_S - 2m(r_S))} \qquad (3.95)$$

Reorganizing terms to obtain an effective interpretation with $dm(r_S) = 4r_S^2(r_S)dr_S$, to clarify the meaning of this equation.

$$\text{Counter Force Acting On The Shell} = 4\pi r_S^2 dP(r_S) \qquad (3.96)$$

$$\text{Newtonian Term} = \frac{m(r_S)dm(r_S)}{r_S^2} \qquad (3.97)$$

$$4\pi r_S^2 dP(r_S) = \frac{m(r_S)dm(r_S)}{r_S^2}\left(1+\frac{P(r_S)}{\rho(r_S)}\right)\left(1+\frac{4\pi r_S^3 P(r_S)}{m(r_S)}\right)\left(1-\frac{2m(r_S)}{r_S}\right)^{-1} \qquad (3.98)$$

The left hand side of the equation shows us what force is exerted by the pressure on a Schwarzschild radius inside the star. If the collapsing star is in equilibrium, this outward pressure is counteracted with the gravitational force of the star, which is the right hand side of the equation. On the right side of the equation, the first term is the Newtonian gravitational attraction, and an additional three terms of correction for relativistic effects. From the first and second terms of correction terms the equation clarifies that is not only mass that causes gravitational attraction, but also pressure, which is not surprising since mass has energy density, and pressure and energy density are equivalent.

Spatiotemporal Pressure ≡ *Energy Density of Mass*

$$P_{ST} \equiv \rho_m \qquad (3.99)$$

In the Tolman-Oppenheimer-Volkoff equation, the numerator is larger than the denominator, than in the term of the Newtonian equation, which means that there is higher pressure inside the star than outside. Moreover, there is a rapid accelerated increase of the pressure due to the pressure correction term in the numerator. This last conclusion indicates that there are finally fundamental limits on masses for strongly gravitating objects, which entails the certitude that the star cannot be in equilibrium if $2m(r_S) > r_S$, since the pressure gradient would increase very rapidly, and the star would go supernova.

Physical or Intangible States with the Tolman-Oppenheimer-Volkoff Equation.

Let us design an ideal representation of a star with radius " R " and a constant density $\rho_0 = 0$. The initial condition of a constant is purely theoretical since stars do not have a constant density. Hence, let us assume the ideal initial condition to assist the clear definition and aim of the physical conceptualization in this context. A constant density insures that the following volume expressions apply to the enclosed stellar mass:

$$m(r_S) = \begin{cases} \dfrac{4}{3}\pi\rho_0 r_S^3 & r_S < R \\ \dfrac{4}{3}\pi\rho_0 R^3 \equiv M & r_S > R \end{cases} \qquad (3.100)$$

The line element, with the definition of *m(r)* in terms of " ε ," may be denoted as,

$$ds^2 = -e^{2\phi}dt^2 + \frac{dr_S^2}{\left(1 - \frac{2m(r_S)}{r_S}\right)} + r_S^2 d\Omega^2 \qquad (3.101)$$

Using the previous equations to study the line element and the hydrostatic properties of the star we have: the equation of the mass, with boundary condition $m(0) = 0$, the Tolman-Oppenheimer-Volkoff equation, with boundary condition $p(R) = 0$, without pressure at the boundary of the star, and the gravitational field equation for "ϕ", with boundary condition $\phi(\infty) = 0$ for nearly flat Minkowski space-time at infinity.

The external medium of the star.

The boundary of the star is defined as the point where "P" becomes zero, and remains zero due to the Tolman-Oppenheimer-Volkoff equation, because at the boundary "ρ" is also zero). Outside the star there is only space-time, so only "ϕ" and the volumetric index "ε" are significant in this context. Integrating the equation for "ϕ", we have

$$\phi(r_S) = \frac{1}{2}\log\left(1 - \frac{2M}{r_S}\right) \qquad (3.102)$$

Hence, in this manner, the geometry reduces to the Schwarzschild metric in accordance with Birkhoff's theorem.

$$ds^2 = -\left(1 - \frac{2M}{r_S}\right)dt^2 + \left(1 - \frac{2M}{r_S}\right)^{-1}dr_S^2 + r_S^2 d\Omega^2 \qquad (3.103)$$

The internal medium of the star.

The Tolman-Oppenheimer-Volkoff equation may be solved after substituting into the $m(r)$ equation for the pressure at the core of the star,

$$P_C(r_S) = \rho_0 \left\{ \frac{\left(1-2Mr_S^2/R^3\right)^{1/2} - \left(1-2M/R\right)^{1/2}}{3\left(1-2M/R\right)^{1/2} - \left(1-2Mr_S^2/R^3\right)^{1/2}} \right\} \quad (3.104)$$

This result is significant, because the physical bounds of the size and mass of the star emerge in a clear and detailed manner. Thus, the pressure "P_C" at the core of the star is given by

$$P_C = P_C(0) = \rho_0 \left\{ \frac{1-\left(1-2M/R\right)^{1/2}}{3\left(1-2M/R\right)^{1/2} - 1} \right\} \quad (3.105)$$

If the denominator equals zero, $(3(1-2M/R)^{1/2} - 1) = 0$, the pressure equation reaches into infinity for a limiting value given by $M/R > 4/9$ or ~0.44, which is also a requirement for the collapsing star, along with $m(r_S/2 > r_S)$. Thus, the star can no longer be in equilibrium, and something must occur, what will occur will be the collapse of the star, as graphically shown in the following figure, where the pressure inside a star is plotted for various mass/radius ratios. For the ratio $M/R < 4/9$, the pressure at the core "Pc" remains finite, and for the ratio $M/R = 4/9$ the pressure scales as $1/r_S^2$, producing an intangible state at the core. In the following figure, it is graphically shown how the core pressure develops as a function of the mass/radius ratio "M/R." The critical value, $M/R = 4/9$, is clearly visible. It is interesting to note how sharp the core pressure increases near the critical value, due to the occurrence of "$P(r_S)$" at the right hand side of the Tolman-Oppenheimer-Volkoff equation. If "$P(r_S)$" is already large, it will accelerate more, because its gradient, on the left hand side of the above equation, will also be large.

The exact meaning of the fundamental limit on the mass of a star has not been straightforward, because one needs an equation of state, relating the pressure to the density inside the star. Moreover, a star will eventually collapse long before the pressure has reached into infinity, because the fermion force of repulsion is finite. The calculation of the Tolman-Oppenheimer-Volkoff limit had been done in detail for neutron stars, where the final result of the calculation was that the mass of a neutron star cannot exceed 0.7 solar masses. That result has been improved to a range of 1.5 to 3.0 due to better knowledge over time on the process of a neutron star formation.

Aside, the ratio of "$\nabla P/\rho_m = -g$" where "g" is gravity. The structure of the neutron star is not known yet, and that introduces some unpredictability. There is the possibility that there might be a quark gluon plasma surrounded by a fermi liquid inside the structure of a collapsing star.

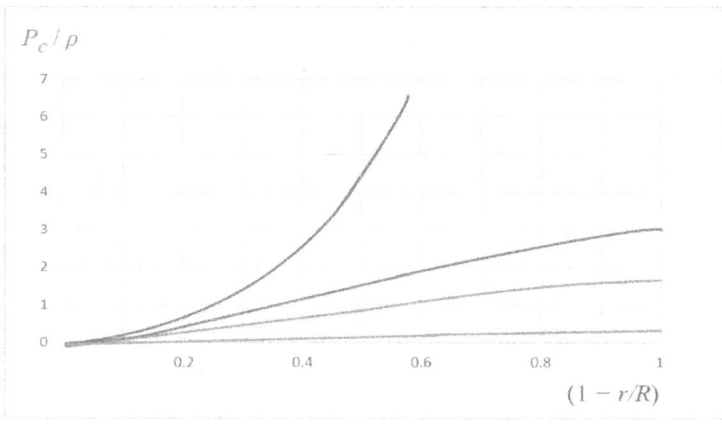

Figure 18. The pressure "P" exerted at a shell at radial position "r" inside the star for various ratios "M/R." The blue curve corresponds to $M/R = 4/9 \approx 0.44$; this value increases the pressure toward infinity at $r/R \approx 0.44$ or $(1- r/R) \approx 0.56$, red curve: $M/R = 0.42$, green curve: $M/R = 0.4$, brown curve: $M/R = 0.3$.

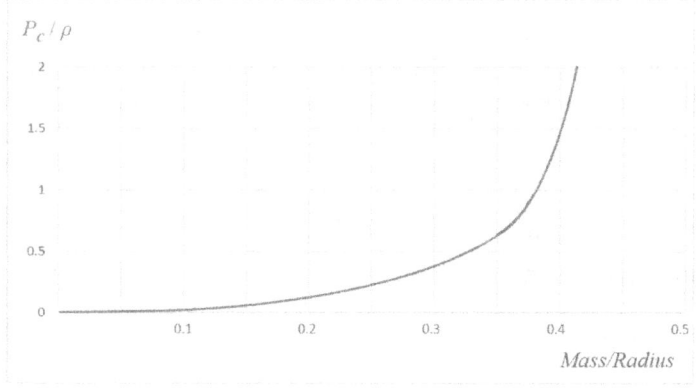

Figure 19. The pressure at the core of the star plotted against the mass/radius ratio, M/R. The fundamental limit of this ratio is $M/R = 4/9$.

Realistic model for a rotating charged star.

Let us consider a more realistic model of a star, taking additional factors into account. First, the internal density of a star will not be constant, there is no fluid that is incompressible. Secondly, a model of a rotating star is significant because most neutron stars rotate very fast. Third, a star that is charged may also be considered. However, charged stars may not be very common because their charges would be neutralized when their charges attract and accrete oppositely charged particles and matter.

There is an extension of the Tolman-Oppenheimer-Volkoff equation for a charged star. The Reissner-Nordström metric may be used outside of the star with the assumption of the relation $g_{00}g_{ii} = -1$, which is also valid for the interior solution of the line element. So, the equation of structure has the following rectification:

$$\frac{dP}{dr_s} = -\frac{(\rho(r_s) + P(r_s))(m(r_s) + 4\pi r_s^3 P(r_s))}{r_s(r_s - 2m(r_s))} + \frac{1}{8\pi r_s^4}\frac{d(q(r_s)^2)}{dr_s} \quad (3.106)$$

with $q(r_s)$ as the enclosed charge. The above equation may be solved for a given charge distribution of charges "$\sigma(r)$"; however, it seems that the total gravitational mass of the star scales with the parameter "$\sigma(r)$". For $m(r) = 0$, and $\sigma = 0$; so, the gravitational mass is entirely of electromagnetic origin, which is not the case for neutron stars that consist of hadrons, or white dwarfs. This solution method may be a method where the Tolman-Oppenheimer-Volkoff equation can be solved in a clear and detailed manner for a charged star.

Solving the Tolman-Oppenheimer-Volkoff equation for the pressure acceleration of a realistic model for a rotating charged star by applying the product rule, $(f \cdot g)' = f' \cdot g + f \cdot g'$, on the following equation, we obtain

$$\frac{dP}{dr_s} = -(\rho(r_s)c^2 + P(r_s))\frac{d\phi}{dr_s} + \frac{1}{8\pi r_s^4}\frac{d(q(r_s)^2)}{dr_s} \quad (3.107)$$

Where $d\phi/dr_S$ and $d^2\phi/dr_S^2$ may not be equal to zero, we get

$$\frac{d^2P}{dr_S^2} = -\left(c^2\frac{d\rho(r_S)}{dr_S} + \frac{dP(r_S)}{dr_S}\right)\frac{d\phi}{dr_S} - \frac{d^2\phi}{dr_S^2}\left(\rho(r_S)c^2 + P(r_S)\right)$$

$$-\frac{1}{8\pi}\left(\frac{4}{r_S^5}\cdot 2q(r_S)\frac{dq(r_S)}{dr_S} - 2\left(\frac{dq(r_S)}{dr_S}\right)^2 + \frac{d^2q(r_S)}{dr_S^2}q(r_S)\right)\frac{1}{r_S^4}\right) \quad (3.108)$$

$$\frac{d^2P}{dr_S^2} = -\frac{d^2\phi}{dr_S^2}\left(\rho(r_S)c^2 + P(r_S)\right) - \frac{d\phi}{dr_S}\left(c^2\frac{d\rho(r_S)}{dr_S} + \frac{dP(r_S)}{dr_S}\right)$$

$$-\frac{-r_S\dfrac{d^2q(r_S)}{dr_S^2}q(r_S) + r_S\left(\dfrac{dq(r_S)}{dr_S}\right)^2 + 4q(r_S)\dfrac{dq(r_S)}{dr_S}}{4\pi r_S^5} \quad (3.109)$$

It is significant to point out that the sign of the charge may increase or decrease the counter gravitational pressure acceleration by adding to, or subtracting from, the effect of the counter spatiotemporal curvature.

It seems that a charge of $\sim 10^{20}$ Coulombs is needed to have any observable effects on the structure of the massive star. If these compact stars that are highly charged do exist, they would probably be isolated stars, because outside the star, the Coulomb force would overcome the force of the gravitational field. The envelope of gases surrounding the charged star would attract unlike charged particles and matter, and repel like charged particles and matter.

Moreover, it seems that in a rotating charged star, the mass can be about twenty percent larger than the mass of a nonrotating charged star. Notwithstanding the gravitational attraction, or the effects of polarity, if two near rotating unlike charged stars were spinning in the opposite direction as they collapse, the spatiotemporal flow would be higher, resulting in lower spatiotemporal pressure, causing mutual spatiotemporal attraction. If the two rotating stars were spinning in the same direction, the two collapsing stars would experience mutual spatiotemporal repulsion.

If a photon pair were spinning near each other similar relativistic effects would be found, creating an attractive or a repulsive effect. Hence, the direction of spin may be considered the polarity of torsion, and the alternating frequency may change the sign, or the direction, of the effect of polarity.

The counter gravitational EFE.

The following counter gravitational EFE is based on the Tolman-Oppenheimer-Volkoff pressure equation.

$$\frac{1}{M_{\mu\nu}} \cdot \frac{c^4 R}{8\pi c^2 G}\left(-C_{i_\mu i_\nu} + \frac{1}{n+1} g_{i_\mu i_\nu} C\right) = T_{\mu\nu} - \Phi_{\mu\nu} \qquad (3.110)$$

Where "$R = g^{\mu\nu} R_{\mu\nu}$" is the trace of the counter gravitational Ricci tensor for the local conjugate spatial curvature, and the left side of the equation represents the spatial acceleration of pressure divided by the stress-mass-momentum tensor $M_{\mu\nu}$. The tensor $C_{i_\mu i_\nu}$ is the counter gravitational tensor and $g_{i_\mu i_\nu}$ is the counter gravitational metric tensor.

$$-C_{i_\mu i_\nu} + \frac{1}{n+1} g_{i_\mu i_\nu} C = \frac{8\pi G}{c^2 R}\left(T_{\mu\nu} - \Phi_{\mu\nu}\right) M_{\mu\nu} \qquad (3.111)$$

The Einstein tensor of temporal curvature, in four-dimensional space-time with folded time, or six-dimensional space-time is given by

$$G_{i_\mu i_\nu} \equiv -C_{i_\mu i_\nu} + \frac{1}{n+1} g_{i_\mu i_\nu} C \qquad (3.112)$$

Where $M_{\mu\nu}$ is a stress-mass-momentum tensor, $T_{\mu\nu}$ is the stress-energy-momentum tensor, $\Phi_{\mu\nu}$ is the electromagnetic tensor for a rotating charged mass, "i_μ" and "i_ν" are the indices $(-1,-2,-3)$ for the three temporal dimensions (t_x, t_y, t_z) equal to "n" in this context of six-dimensional space-time; so, $n = -3$ for the above EFE.

$$M_{\mu\nu} = \frac{\rho_{i_\mu i_\nu} c^2 + P_{\mu\nu}}{c^2} = \rho_m + \rho_P \quad (3.113)$$

Hence, in the counter gravitational EFE, it is interesting to note that the mass density is the source of the temporal curvature while the energy density is the source of the conjugate spatial curvature.

Solving for the Tolman-Oppenheimer-Volkoff equation for the pressure acceleration assuming the ideal initial condition of a constant density:

$$\frac{dP}{dr_S} = -\frac{(\rho+P)(m+4\pi r_S^3 P)}{r_S(r_S-2m)} \quad (3.114)$$

Applying the quotient rule, $\left(\frac{f}{g}\right)' = \frac{f' \cdot g - g' \cdot f}{g^2}$, on the numerator, we have

$$-\frac{d^2 P}{dr_S^2} = -\frac{d^2\left(\frac{(\rho+P)(m+4\pi r_S^3 P)}{r_S(r_S-2m)}\right)}{dr_S^2} = -\frac{1}{(r_S^2+2mr_S)^2}\left(8\pi r_S^5 P \frac{dP}{dr_S} + 4\pi\rho r_S^5 \frac{dP}{dr_S} + \right.$$

$$4\pi r_S^4 P^2 + 16\pi m r_S^4 P \frac{dP}{dr_S} + 8\pi m\rho r_S^4 \frac{dP}{dr_S} + 4\pi\rho r_S^4 P + 16\pi m r_S^3 P^2 +$$

$$16\pi m\rho r_S^3 P + m r_S^2 \frac{dP}{dr_S} + 2m^2 r_S \frac{dP}{dr_S} - 2m r_S P - 2m\rho r_S - 2m^2 P - 2m^2 \rho \quad (3.115)$$

$$-\frac{d^2 P}{dr_S^2} = (\rho c^2 + P)\frac{d^2\phi}{dr_S^2} \quad (3.116)$$

3.5. The Jacobian of spherical coordinates.

In the Cartesian coordinate system, the location of a point in space is described using an ordered triple in which each coordinate represents a distance. In the cylindrical coordinate system, location of a point in space is described using two distances "r" and "z" and an angle measure "θ". In the spherical coordinate system, we again use an ordered triple to describe the location of a point in space. In this case, the triple describes one distance and two angles. Spherical coordinates make it simple to describe a sphere, just as cylindrical coordinates make it easy to describe a cylinder. Grid lines for spherical coordinates are based on angle measures, like those for polar coordinates.

Spherical or Cartesian coordinates give us the flexibility to select a coordinate system appropriate to the problem at hand, like to define curvature and torsion or volumetric divergence or convergence, for a tesseract. A thoughtful choice of coordinate system can make a problem much easier to solve, whereas a poor choice can lead to unnecessarily complex calculations.

What is the relation between Cartesian, spherical and cylindrical coordinates for the Jacobian?

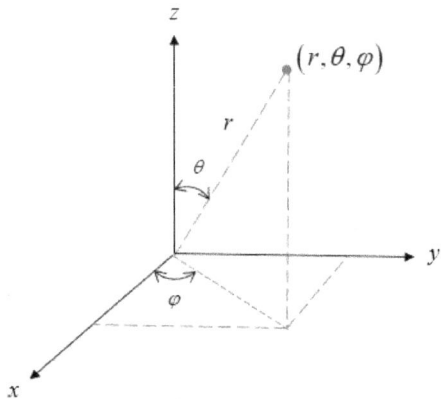

Figure 20. The Relationship between Spherical and Cartesian Coordinates as commonly used in Physics.

Spherical coordinates of the system denoted as (ct_p, ϕ_n, ω_n) is the coordinate system used in six-dimensional spatiotemporal systems with folded time. In six-dimensional space-time with folded time, the spherical coordinate system is used for finding the surface area.

These coordinates specify three numbers: radial distance "ct_p," polar angle "ϕ_n" and azimuthal angle "ω_n."

In the spherical coordinate system, a point "P" in space is represented by the ordered triple (ct_p, ϕ_n, ω_n) where "ct_p" (the speed of light times the Planck time) is the distance between "P" and the origin "O," the angle "ϕ_n" is used to describe the location from the z-axis to the radial distance "ct_p;" the angle "ω_n" is formed by the positive x-axis and the projection of the line segment "OP" on the xy-plane, where "O" is the origin and "P" is a point, and $0 \leq \omega_n \leq 2\pi$.

The relations between Cartesian and Spherical coordinates are:

$$x = ct_p \operatorname{Sin}(\phi_n) \operatorname{Cos}(\omega_p) \qquad (3.117)$$

$$y = ct_p \operatorname{Sin}(\phi_n) \operatorname{Sin}(\omega_p) \qquad (3.118)$$

$$z = ct_p \operatorname{Cos} \phi_n \qquad (3.119)$$

The following equations represent the relation between Cartesian coordinates to Spherical coordinates for a point:

$$c^2 t_p^2 = x^2 + y^2 + z^2 \qquad (3.120)$$

$$\tan \omega_n = \frac{y}{x} \qquad (3.121)$$

$$\phi_n = \frac{arc \operatorname{Cos} z}{x^2 + y^2 + z^2} \qquad (3.122)$$

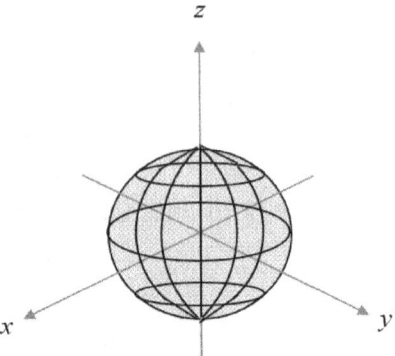

Figure 21. The Relationship between the Two-Dimensional Jacobian and the Three-Dimensional Jacobian for a Sphere.

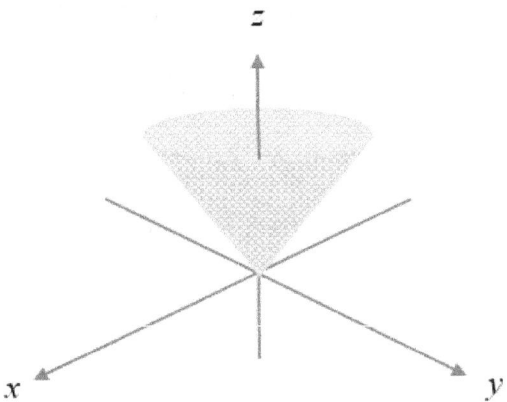

Figure 22. The Relationship between the Two-Dimensional Jacobian and the Three-Dimensional Jacobian for a Cone.

The geographical latitude and longitude are used to describe locations on the surface of the Earth. In the system of latitude and longitude, the angles describe the location of a point on the surface of the Earth relative to the equator and the prime meridian. Let us imagine that the Earth has the shape of a sphere with radius "r," not the actual shape of an oblate spheroid.

Angles measure distances in degrees rather than radians because latitude and longitude are measured in degrees. The prime meridian represents the trace of the surface as it intersects the xz-plane. The equator is the trace of the sphere intersecting the xy-plane.

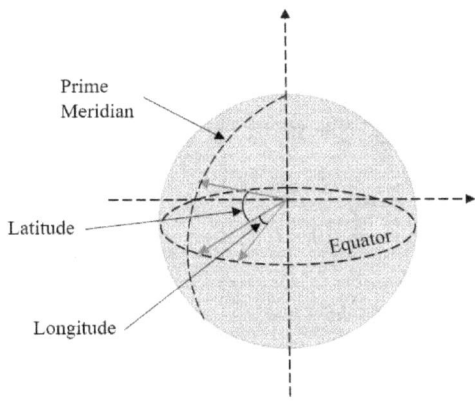

Figure 23. An Illustration of the Geographical Latitude and Longitude.

3.6. The Jacobian spatiotemporal scale factor for a tesseract.

The spatiotemporal deformation factor between size in uv-space-time and size in xy-space-time is called the Jacobian. The Jacobian of the transformation is found by taking the determinant of a two by two matrix of partial derivatives. The Jacobian matrix is used to measure the spatiotemporal curvature and torsion of a tesseract, hypercube, or a hypocube in six-dimensional space-time or in four-dimensional space-time with folded time. The value of the Jacobian varies from −1 to +1. The Jacobian ratio measures the deviation of a line element's form from an ideally formed element, a line element that has straight edges with equal lengths. Furthermore, the Jacobian ratio of a perfect second order tetrahedral element with linear edges is +1.0. The ideal value of Jacobian is +1 and it decreases as the measured line element becomes more deformed.

$$\text{Scale Factor} = J\left(t_p, \omega_n\right) \quad (3.123)$$

$$\text{Scale Factor} \equiv \frac{\text{Area of (Radius of Curvature} \times \text{Angle of Torsion)}}{\text{Area of (Unit Radius of Curvature} \times \text{Unit Angle of Torsion)}} \quad (3.124)$$

$$\text{Scale Factor} \equiv \frac{\text{Area of Curvature} \cdot \text{Torsion}}{\text{Unit Area of Curvature} \cdot \text{Torsion}} \quad (3.125)$$

$$J(t_p, \omega_p) \equiv \int_0^{2\pi} \int_0^{t_p} dt_p \, d\omega_n \quad (3.126)$$

$$\text{Jacobian Matrix} = \begin{bmatrix} \dfrac{\partial t_p}{\partial t_p} \text{Cos}(\omega_n) & \dfrac{\partial t_p}{\partial \omega_n} \text{Cos}(\omega_n) \\ \dfrac{\partial t_p}{\partial t_p} \text{Sin}(\omega_n) & \dfrac{\partial t_p}{\partial \omega_n} \text{Sin}(\omega_n) \end{bmatrix} \quad (3.127)$$

The Jacobian determinant:

Let us consider a small area "dA" to denote a rectangular area spanned by "dx" and "dy," then "dA" approximates the area of the scalar curvature of the pseudo-Riemannian manifold for "du" and "dv" sufficiently close to 0. That is, the area of an infinitesimal region in the uv-manifold is scaled by the Jacobian determinant to approximate an area of an infinitesimal region in the xy-manifold. A spatiotemporal manifold is a topological space-time that locally may resemble Euclidean space-time near each point.

$$J(t_p, \omega_n) \partial t_p \partial \omega_p \rightleftarrows \partial x \partial y \quad (3.128)$$

$$J(t_p, \omega_n) \rightleftarrows \dfrac{\partial x \partial y}{\partial t_p \partial \omega_p} \quad (3.129)$$

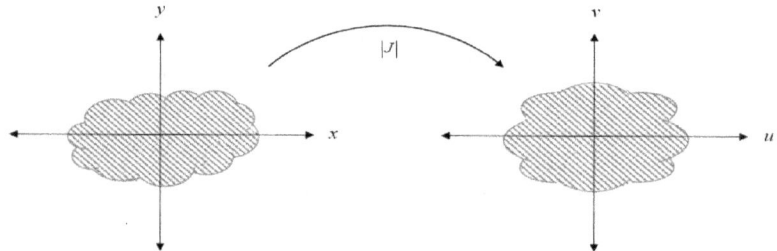

Figure 24. An Illustration of the xy-space-time and the uv-space-time coordinates used to find the Jacobian's determinant.

Where $x = \dfrac{\partial t_p}{\partial t_p} \text{Cos}(\omega_p)$ and $y = \dfrac{\partial t_p}{\partial t_p} \text{Sin}(\omega_p)$. \quad (3.130)

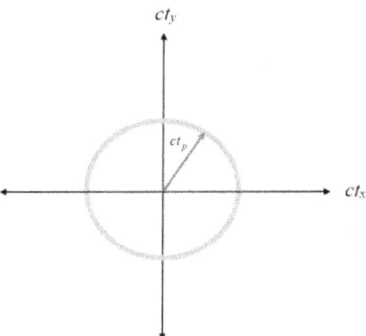

Figure 25. The *xy*-space-time coordinate system showing the radius "t_p."

Computing the scale factor from the determinant of the Jacobian matrix,

$$\det \begin{bmatrix} \frac{\partial t_p}{\partial t_p} \cos(\omega_n) & -\frac{\partial t_p}{\partial \omega_n} \sin(\omega_n) \\ \frac{\partial t_p}{\partial t_p} \sin(\omega_n) & \frac{\partial t_p}{\partial \omega_n} \cos(\omega_n) \end{bmatrix} = \frac{\partial(x,y)}{\partial(t_p,\omega_n)} \quad (3.131)$$

$$\det \begin{bmatrix} \frac{\partial t_p}{\partial t_p} \cos(\omega_n) & -\frac{\partial t_p}{\partial \omega_n} \sin(\omega_n) \\ \frac{\partial t_p}{\partial t_p} \sin(\omega_n) & \frac{\partial t_p}{\partial \omega_n} \cos(\omega_n) \end{bmatrix} = \left| \frac{\partial t_p}{\partial \omega_n} \cdot \cos^2(\omega_n) + \frac{\partial t_p}{\partial \omega_n} \cdot \sin^2(\omega_n) \right| = \frac{\partial t_p}{\partial \omega_n} \quad (3.132)$$

The Jacobian represents the best uncurved, or linear, approximation to a differentiable function near a given spatiotemporal point. In this context, the Jacobian is the derivative of a spatiotemporal multivariate function, and is itself a tensor. The spatiotemporal gradients "$\vec{\Re}$" for six-dimensional space-time or "$\vec{\nabla}$" for three-dimensional space are subsets of the Jacobian. (Nieves, 2020) The Jacobian of a scalar function is the transpose of its gradient. The Jacobian is the determinant of the Jacobian matrix. The Jacobian matrix will contain all partial derivatives of a vector function. The transformation of coordinates is the main use of the Jacobian. The Jacobian employs the concept of differentiation with coordinate transformation.

In the finite line element method, a line element's Jacobian matrix relates the quantities given in the spatiotemporal coordinate system. The more the line element is distorted in comparison with an ideal form line element, the more deformed will be the transformation of the quantities from the initial spatiotemporal coordinate system to the actual spatiotemporal coordinate system. If the Jacobian is zero, it means that there is no change whatsoever in the curvature and torsion, or in the spatiotemporal deformation, of the trajectory from flat space-time (Minkowski's space-time), or from the initial curved or nearly flat space-time. to the final curved space-time, or final nearly flat space-time. Therefore, this means that one gets an overall change of zero at that point, with respect to the rate of change with respect to the spatiotemporal expansion and/or contraction, in relation to the whole spatiotemporal volume.

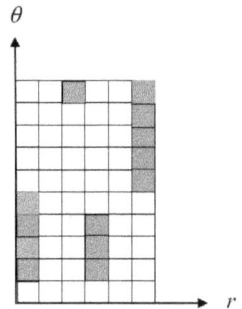

Figure 26. An illustration of a Cartesian coordinate plane.

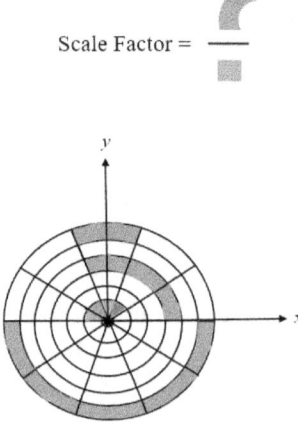

Figure 27. An illustration of a Spherical coordinate circular surface.

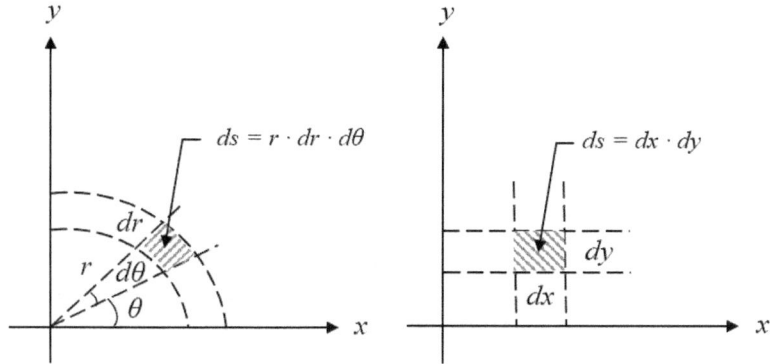

Figure 28. Differentials of Surface.

3.7. The Jacobian for the volumetric transformation of spatiotemporal coordinates.

In the previous section, multivariable functions were defined in terms of Cartesian coordinates "x" and "y" were converted into functions defined in terms of polar coordinates "ℓ_p" and "ω_n."

Similarly, given a region defined in uvw-space-time, we can use a Jacobian transformation to redefine it in xyz-space-time, or vice versa. In any coordinate system, in computing an integral over a volume, you break the volume up into little pieces, and sum the value of the integrand at a point in each piece, times the volume of the piece.

The Jacobian tells you how to express the volume element $\partial x \partial y \partial z$ in the new coordinates $\partial u \partial v \partial w$.

The Jacobian of Spherical Coordinates is given by

$$J = \begin{vmatrix} \dfrac{1}{c}\dfrac{\partial x}{\partial t_p} & \dfrac{\partial x}{\partial \phi_n} & \dfrac{\partial x}{\partial \omega_n} \\ \dfrac{1}{c}\dfrac{\partial y}{\partial t_p} & \dfrac{\partial y}{\partial \phi_n} & \dfrac{\partial y}{\partial \omega_n} \\ \dfrac{1}{c}\dfrac{\partial z}{\partial t_p} & \dfrac{\partial z}{\partial \phi_n} & \dfrac{\partial z}{\partial \omega_n} \end{vmatrix} = \begin{vmatrix} \cos\phi_n \sin\omega_n & -ct_p \sin\phi_n \sin\omega_n & ct_p \cos\phi_n \cos\omega_n \\ \sin\phi_n \sin\omega_n & ct_p \cos\phi_n \sin\omega_n & ct_p \sin\phi_n \cos\omega_n \\ \cos\phi_n & 0 & -ct_p \sin\phi_n \end{vmatrix} \quad (3.133)$$

$$= -c^2t_p^2 \cos^2\phi_n \sin^2\omega_n - c^2t_p^2 \sin^2\phi_n \cos^2\omega_n \sin\omega_n - c^2t_p^2 \cos^2\phi_n \cos^2\omega_n \sin\omega_n \quad (3.134)$$

$$-c^2t_p^2 \sin^2\phi_n \sin^3\omega_n = -c^2t_p^2 \sin^3\omega_n \left(\cos^2\phi_n \sin^2\phi_n\right) - c^2t_p^2 \cos^2\omega_n \sin\omega_n \left(\cos^2\phi_n + \sin^2\phi_n\right) \quad (3.135)$$

$$= -c^2t_p^2 \sin^3\omega_n - c^2t_p^2 \cos^2\omega_n \sin\omega_n = -c^2t_p^2 \sin\omega_n \left(\sin^2\omega_n + \cos^2\omega_n\right) = -c^2t_p^2 \sin\omega_n \quad (3.136)$$

$$J = -c^2t_p^2 \sin\omega_n \quad (3.137)$$

Therefore, the Jacobian of Spherical coordinates may not be zero, unless the angle of torsion is 0°, 180°, or 360°, because the $\sin\omega_n$ term of the spatiotemporal wave of the Jacobian at those angles is zero.

Hence, 0° < ω_n < 180° or 180° < ω_n < 360° for either divergence or convergence to emerge, but at each of the angles 0°, 180°, or 360°, the symmetrical tesseract returns to its initial spatiotemporal volume, so the change in spatiotemporal volume is zero.

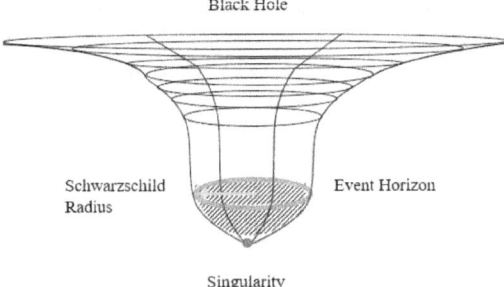

Figure 29. The integration region of a Schwarzschild black hole, with $(x^2 + y^2 = n \cdot z)$ for the cone that intersects the plane $(z = n)$, where "n" is the value of a "z" coordinate.

Consequently,

$$\partial x \partial y \partial z = |J| c \partial t_p \partial \phi_n \partial \omega_n = c^2 \partial t_p^2 \left(\sin\omega_n\right) c \partial t_p \partial \phi_n \partial \omega_n \quad (3.138)$$

For the instance of ($n = 3$) spatiotemporal variables, given three functions $x = f\left(ct_p, \phi_n, \omega_n\right)$, $y = g\left(ct_p, \phi_n, \omega_n\right)$, and $z = h\left(ct_p, \phi_n, \omega_n\right)$, the Jacobian takes the form,

$$J f\left(ct_p, \phi_n, \omega_n\right) \equiv \left| \frac{1}{c} \frac{\partial P(x,y,z)}{\partial t_p} \cdot \frac{\partial P(x,y,z)}{\partial \phi_n} \cdot \frac{\partial P(x,y,z)}{\partial \omega_n} \right| \quad (3.139)$$

Where the Jacobian represents the divergence or convergence of a spatiotemporal volume at a point "P" defined in terms of Cartesian and Spherical coordinates.

$$J = \left| \frac{\partial P(x,y,z)}{\partial P(ct_p, \phi_n, \omega_n)} \right| = \begin{vmatrix} \frac{1}{c}\frac{\partial x}{\partial t_p} & \frac{\partial x}{\partial \phi_n} & \frac{\partial x}{\partial \omega_n} \\ \frac{1}{c}\frac{\partial y}{\partial t_p} & \frac{\partial y}{\partial \phi_n} & \frac{\partial y}{\partial \omega_n} \\ \frac{1}{c}\frac{\partial z}{\partial t_p} & \frac{\partial z}{\partial \phi_n} & \frac{\partial z}{\partial \omega_n} \end{vmatrix} \quad (3.140)$$

If there is no divergence or convergence, the Jacobian would be as follows:

$$J = \left| \frac{\partial P(x,y,z)}{\partial P(ct_p, \phi_n, \omega_n)} \right| = \begin{vmatrix} \frac{1}{c}\frac{\partial x}{\partial t_p} & \frac{\partial x}{\partial \phi_n} & 0 \\ 0 & \frac{\partial y}{\partial \phi_n} & \frac{\partial y}{\partial \omega_n} \\ \frac{1}{c}\frac{\partial z}{\partial t_p} & 0 & \frac{\partial z}{\partial \omega_n} \end{vmatrix} = \begin{vmatrix} \frac{1}{\phi_n} & \frac{-ct_p}{\phi_n^2} & 0 \\ 0 & \frac{1}{\omega_n} & -\frac{\phi_n}{\omega_n^2} \\ -\frac{\omega_n}{c^2 t_p^2} & 0 & \frac{1}{ct_p} \end{vmatrix} \quad (3.141)$$

$$= \frac{1}{c}\frac{\partial x}{\partial t_p} \begin{vmatrix} \frac{\partial y}{\partial \phi_n} & \frac{\partial y}{\partial \omega_n} \\ \frac{\partial z}{\partial \phi_n} & \frac{\partial z}{\partial \omega_n} \end{vmatrix} - \frac{\partial x}{\partial \phi_n} \begin{vmatrix} \frac{\partial y}{\partial t_p} & \frac{\partial y}{\partial \omega_n} \\ \frac{\partial z}{\partial t_p} & \frac{\partial z}{\partial \omega_n} \end{vmatrix} + \frac{\partial x}{\partial \omega_n} \begin{vmatrix} \frac{\partial y}{\partial t_p} & \frac{\partial y}{\partial \phi_n} \\ \frac{\partial z}{\partial t_p} & \frac{\partial z}{\partial \phi_n} \end{vmatrix} \quad (3.142)$$

$$= -\frac{1}{c}\frac{\partial x}{\partial t_p}\left(\frac{\partial y}{\partial \phi_n}\cdot\frac{\partial z}{\partial \omega_n} - \frac{\partial y}{\partial \omega_n}\cdot\frac{\partial z}{\partial \phi_n}\right) - \frac{\partial x}{\partial \phi_n}\left(\frac{\partial y}{\partial t_p}\cdot\frac{\partial z}{\partial \omega_n} - \frac{\partial y}{\partial \omega_n}\cdot\frac{\partial z}{\partial t_p}\right) + \frac{\partial x}{\partial \omega_n}\left(\frac{\partial y}{\partial t_p}\cdot\frac{\partial z}{\partial \phi_n} - \frac{\partial y}{\partial \phi_n}\cdot\frac{\partial z}{\partial t_p}\right) \quad (3.143)$$

$$J = \begin{vmatrix} \frac{1}{\phi_n} & \frac{-ct_p}{\phi_n^2} & 0 \\ 0 & \frac{1}{\omega_n} & -\frac{\phi_n}{\omega_n^2} \\ -\frac{\omega_n}{c^2 t_p^2} & 0 & \frac{1}{ct_p} \end{vmatrix} = \frac{1}{\phi_n}\begin{vmatrix} \frac{1}{\omega_n} & -\frac{\phi_n}{\omega_n^2} \\ 0 & \frac{1}{ct_p} \end{vmatrix} - \left(\frac{-ct_p}{\phi_n^2}\right)\begin{vmatrix} 0 & -\frac{\phi_n}{\omega_n^2} \\ -\frac{\omega_n}{c^2 t_p^2} & \frac{1}{ct_p} \end{vmatrix} + 0\begin{vmatrix} 0 & \frac{1}{\omega_n} \\ -\frac{\omega_n}{c^2 t_p^2} & \frac{1}{ct_p} \end{vmatrix} \quad (3.144)$$

$$J = \frac{1}{\phi_n}\left[\frac{1}{\omega_n}\cdot\frac{1}{ct_p} - \left(-\frac{\phi_n}{\omega_n^2}\cdot 0\right)\right] + \frac{ct_p}{\phi_n^2}\left[0\cdot\frac{1}{ct_p} - \left(-\frac{\phi_n}{\omega_n^2}\cdot -\frac{\omega_n}{c^2 t_p^2}\right)\right] \quad (3.145)$$

Simplifying and rearranging terms, we have

$$J = \frac{1}{ct_p\cdot\phi_n\cdot\omega_n} - \frac{1}{ct_p\cdot\phi_n\cdot\omega_n} \quad (3.146)$$

$$J = 0 \quad (3.147)$$

When the Jacobian derivative method calculates the divergence or convergence contribution to the energy rate of deformation line element by line element.

Consequently, the contributions are added to the energy rate of deformation of all the elements that surround one sector "$ct_p\cdot\omega_n$" of the pseudo-Riemannian manifold at a time, divide this value by the length of the effect covered by the corresponding sector to obtain the distribution of the energy rate of deformation along the whole surface.

The rows of the Jacobian matrix can also be split into two parts. The three rows are associated with linear velocities of each Cartesian coordinate with respect to each related spherical coordinate, and the three columns are associated with the radial temporal velocity of each Cartesian coordinate and the angular velocities of each Cartesian coordinate due to a change in the torsion of all the radial line elements in a sector "$ct_p \cdot \omega_n$."

The angular velocity "ω" is a pseudo vector given by the product of the axis of rotation "$\hat{\omega}$" and the rate of rotation "$\dot{\omega}_n$" about the axis.

$$\omega = \hat{\omega} \cdot \dot{\omega}_n \qquad (3.148)$$

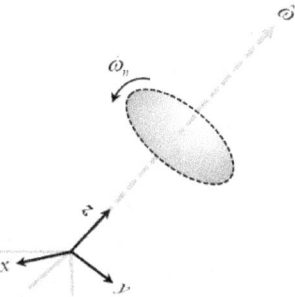

Figure 30. A rotating pseudo-Riemannian manifold about the axis "$\hat{\omega}$" with a velocity "$\dot{\omega}_n$" in rads/sec.

A unit vector "$\hat{\omega}$" represents the axis of rotation in three-dimensional space-time. The unit vector for the rotating pseudo-Riemannian manifold is denoted as

$$\hat{\omega} = a_x \hat{x} + a_y \hat{y} + a_z \hat{z} \qquad (3.149)$$

So, this unit vector may be represented as a (3×1) matrix:

$$\hat{\omega} = \begin{bmatrix} a_x \\ a_y \\ a_z \end{bmatrix} \qquad (3.150)$$

Hence, the angular velocity may be represented in matrix form as follows:

$$\text{Angular Velocity } "\omega" = \begin{bmatrix} a_x \\ a_y \\ a_z \end{bmatrix} \cdot \dot{\omega}_n \qquad (3.151)$$

3.8. Tic-Tac-Zero time at the event horizon.

Let us imagine the first probe that is sent to explore the interior of a supermassive black hole. Inside the probe there is a movie that starts playing continuously before the probe enters in the gravitational field of the black hole, while the same movie will play later for comparison at the launch control center.

An artificial intelligence observer monitors and records the probe's transmitted images as they arrive at the launch control center. The probe was built very rugged to withstand the tidal forces that it may encounter near the black hole.

The onboard commander is a state-of-the-art artificial intelligence unit. The probe's movie recorded in proper time at the launch control center is measured by the probe's onboard clock, through the trajectory in the gravitational field, until the probe crosses the event horizon of the black hole.

As the probe gets closer to the black hole, its transmitted image gets gradually slower, as the probe enters the strong gravitational field near the black hole, the probe gets closer and closer to the event horizon and the transmitted image starts getting very slow, proper time dilates even more.

The probe crosses the event horizon without any particular incident, but the transmitted image freezes, time is at tic-tac-zero on the event horizon, and as soon as the transmitted image arrives at the launch control center on Earth, it suddenly disappears as the transmission ends.

The Einstein effect produces a shift in the frequencies in the strong gravitational field of the black hole, causing the images to weaken, and they rapidly become invisible.

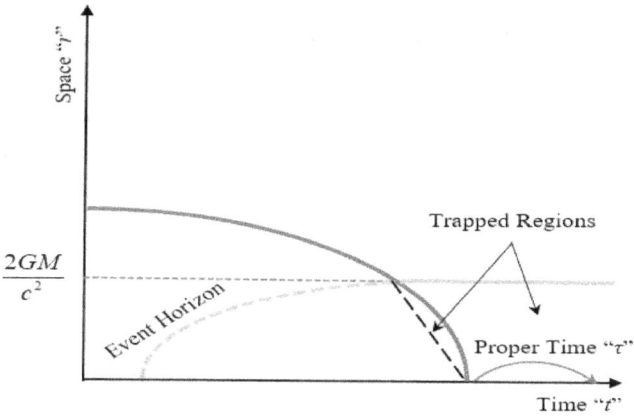

Figure 31. The Apparent and Proper Time of a Black Hole. Adapted from (Straumann, 2019).

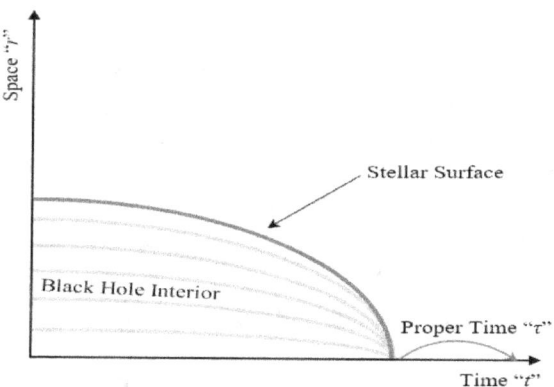

Figure 32. The worldlines of dust shells in an "Oppenheimer-Snyder Star Collapse" reach the singularity at the center of the black hole at the same proper time. Adapted from (Straumann, 2019).

The apparent horizon is located in the outer boundary of the trapped region. Whenever an apparent horizon has been formed, the event horizon can be found on, or rigorously outside, the apparent horizon.

3.9. The spatiotemporal deformation potential of divergence or convergence.

"Everything in physics may be about whether something is spinning, oscillating, traveling on a curve, expanding, contracting, or standing still."

Force × *(Curvature + Torsion + Volumetric Deformation)* ≡
Energy Density

The Jacobian of spherical coordinates may be used to assess the spatiotemporal deformation between a cone and a sphere at every corner of a hypercube or tesseract.

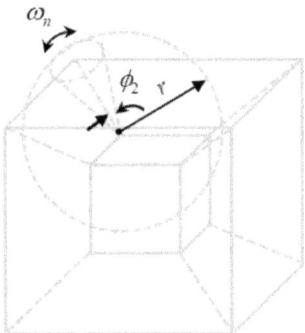

Figure 33. An illustration of a "light cone-Schwarzschild sphere" arrangement on each corner of a hypercube, or tesseract, to gauge spatiotemporal deformation.

The relation between Cartesian and spherical coordinates was given in the Jacobian for spherical coordinates as $J = -c^2 t_p^2 \sin \omega_n$.

Therefore, using the Jacobian volume, we have

$$\partial x \partial y \partial z = |J| c \partial t_p \partial \phi_n \partial \omega_n = c^2 \partial t_p^2 (\sin \omega_n) c \partial t_p \partial \phi_n \partial \omega_n \quad (3.152)$$

Let us consider the integrals where we will use Cartesian, cylindrical, and spherical coordinates: Using cylindrical coordinates to evaluate the following integral in Cartesian coordinates:

$$\iiint_R (x^2 + y^2) dx dy dz \quad (3.153)$$

where "R" is a volume bounded by the surface $x^2 + y^2 = z^2$ and the plane $z = \ell_p \cos \phi_n$. Let us sketch the integration region. The first equation $x^2 + y^2 = z^2$ is a cone, a volume that was connected to each of the corners of the hypercube to describe curvature and torsion. The integration region looks like this:

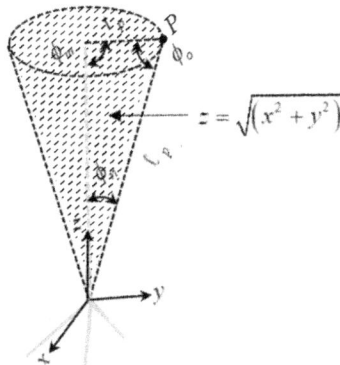

Figure 34. The integration region of a cone.

Now we want to do the integral in cylindrical coordinates. We have seen before that

$$x^2 + y^2 = r^2 \tag{3.154}$$

$$dxdydz = rdrd\theta dz \tag{3.155}$$

so the integral we want to compute is

$$\iiint_{R'} r^3 dr d\theta dz \tag{3.156}$$

where "R'" is the integration region in cylindrical coordinates. From the figure it is possible to find

$$R' = (r, \theta, z) \to \left(0 \le r \le \sqrt{z \ell_p \cos\phi_n}\right), (0 \le \theta \le 2\pi), \left(0 \le z \le \ell_p \cos\phi_n\right) \tag{3.157}$$

Hence, the volumetric integrals are

$$\int_{z=0}^{z=\ell_p \cos\phi_n} dz \int_{\theta=0}^{\theta=2\pi} d\theta \int_{r=0}^{r=\sqrt{z\ell_p \cos\phi_n}} r^3 dr \tag{3.158}$$

The integral in "r" is

201

$$\int_{r=0}^{r=z} r^3 dr = \left[\frac{r^4}{4}\right]_0^{\sqrt{z\ell_p \text{Cos}\phi_n}} = \frac{z^2 \ell_p^2 \text{Cos}^2 \phi_n}{4} \qquad (3.159)$$

The integral in "θ" is

$$\int_{\theta=0}^{\theta=2\pi} d\theta = [\theta]_0^{2\pi} = 2\pi \qquad (3.160)$$

Thus, the result is

$$2\pi \int_{z=0}^{z=\ell_p \text{Cos}\phi_n} z^2 dz = \left[\frac{z^3}{3}\right]_0^{\ell_p \text{Cos}\phi_n} = \frac{2\pi \ell_p^3 \text{Cos}^3 \phi_n}{3} \qquad (3.161)$$

Using spherical coordinates to calculate the spatiotemporal volume bounded above by the sphere $x^2 + y^2 + z^2 = r_s^2$ and below by the cone $z = \sqrt{x^2 + y^2}$. The radius of the sphere "r_s" is the Schwarzschild radius.

The region of integration for the Schwarzschild volume is given in the following figure:

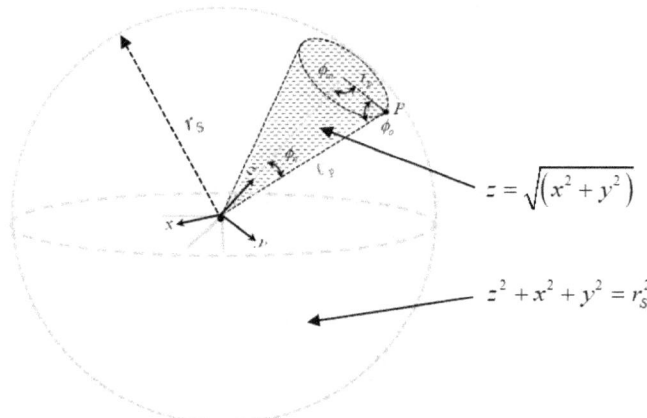

Figure 35. The Schwarzschild sphere $x^2 + y^2 + z^2 = r_s^2$ and the cone $z = \sqrt{x^2 + y^2}$. The dashed region is the integration region.

The integral of the volume is given by

$$\iiint_R dxdydz = \iiint_{R'} r^2 \sin\phi\, dr d\theta d\phi \qquad (3.162)$$

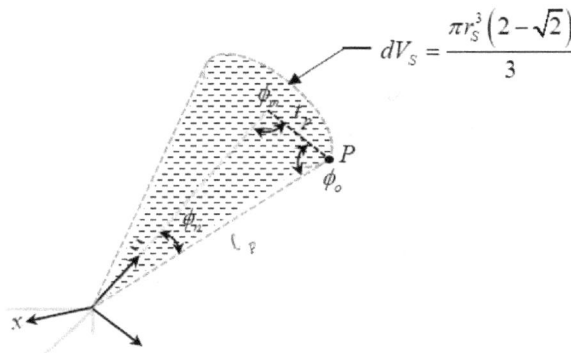

$$dV_S = \frac{\pi r_S^3 (2-\sqrt{2})}{3}$$

Figure 36. The volume of deformation.

Thus, using the Jacobian volume and the integration in spherical coordinates to determine "R'," it is interesting to reiterate again that the equation of the Schwarzschild sphere in spherical coordinates is

$$x^2 + y^2 + z^2 = r_S^2 \qquad (3.163)$$

and the equation of the cone is

$$r\cos\phi = \sqrt{r^2 \sin^2\phi \cos^2\theta + r^2 \sin^2\phi \sin^2\theta} = r\sin\phi \qquad (3.164)$$

From the equation of the cone, we obtain

$$\phi = \frac{\pi}{4} = 45° \quad \rightarrow \quad \tan\phi = 1 \qquad (3.165)$$

With respect to the z-axis, these are the angles of the cone, in order to integrate the region "R'":

$$R' = (r, \theta, \phi) \rightarrow (0 \le r \le r_S),\ (0 \le \theta \le 2\pi),\ (0 \le \phi \le \pi/4) \qquad (3.166)$$

Thus, the volume is

$$dV_S = \int_{r=0}^{r=r_s} r^2 dr \int_{\theta=0}^{\theta=2\pi} d\theta \int_{\phi=0}^{\phi=\pi/4} \sin\phi \, d\phi \quad (3.167)$$

evaluating each integral of the volume

$$\int_{\phi=0}^{\phi=\pi/4} \sin\phi \, d\phi = [-\cos\phi]_{\phi=0}^{\phi=\pi/4} = -\frac{1}{\sqrt{2}} + 1 = \frac{2-\sqrt{2}}{2} \quad (3.168)$$

$$\int_{\theta=0}^{\theta=2\pi} d\theta = [\theta]_0^{2\pi} = 2\pi \quad (3.169)$$

$$\int_{r=0}^{r=r_s} r^2 dr = \left[\frac{r^3}{3}\right]_0^{r_s} = \frac{r_s^3}{3} \quad (3.170)$$

Therefore, the volume of deformation is given by

$$dV_S = \frac{\pi r_s^3 (2-\sqrt{2})}{3} \quad (3.171)$$

Using the Jacobian that represents the potential deformation of divergence or convergence of the Schwarzschild spatiotemporal spherical volume bounded by a light cone over time, we have

$$dV_S^{future} - dV_S^{present} = \frac{\pi r_{S_{future}}^3 (2-\sqrt{2})}{3} - \frac{\pi r_{S_{present}}^3 (2-\sqrt{2})}{3} = \frac{\pi(2-\sqrt{2})}{3}\left(r_{S_{future}}^3 - r_{S_{present}}^3\right) \quad (3.172)$$

$$dV_S^{future} - dV_S^{present} \approx 0.613434123 \left(r_{S_{future}}^3 - r_{S_{present}}^3\right) \quad (3.173)$$

As it was demonstrated previously, if the Schwarzschild radii stay symmetrical, the final result is zero deformation. The symmetrical condition of the Schwarzschild radii results from the interference of spatiotemporal waves, in either four-dimensional space-time with folded time, or six-dimensional space-time, that offset each other resulting in no significant divergence, or convergence, within or very near the Schwarzschild volume under consideration. (Nieves, 2020) Therefore, the curvature, torsion, and deformation given by the Schwarzschild spatiotemporal gauge may be denoted as

$$\Re = \left(\phi_{S_{future}} - \phi_{S_{present}}\right) + \left(\phi_{SC_{future}} - \phi_{SC_{present}}\right) + \frac{\pi\left(2-\sqrt{2}\right)\left(r_{S_{future}}^3 - r_{S_{present}}^3\right) \cdot 180°}{3\left(r_{S_{present}}^3\right) \cdot \pi} \quad (3.174)$$

$$\Re = \left(\phi_{S_{future}} - \phi_{S_{present}}\right) + \left(\phi_{SC_{future}} - \phi_{SC_{present}}\right) + \frac{60°\left(2-\sqrt{2}\right)\left(r_{S_{future}}^3 - r_{S_{present}}^3\right)}{r_{S_{present}}^3} \quad (3.175)$$

The curvature, torsion, and volume deformation is: $\Re \sim d\phi_S + d\phi_{SC} + (3/5)dV_S$. Therefore, the Schwarzschild-Cartan-Jacobian metric for the curvature, torsion, and volume deformation is given by

$$ds^2 = -\left(1 - \phi_S - \phi_{SC} - 60°\left|c^3 t_S^3\right|\left(2-\sqrt{2}\right)\right)dt^2 + \left(1 - \phi_S - \phi_{SC} - 60°\left|r_S^3\right|\left(2-\sqrt{2}\right)\right)^{-1} dr^2 \quad (3.176)$$

References

1. Abedi, Jahed et al. (2021) GW190521: First Measurement of Stimulated Hawking Radiation from Black Holes, DOI: arXiv:2201.00047.

2. Biswas, Ripan*, Sahadath, Hossain*, Mollah, Abdus Sattar**, Huq, Md. Fazlul*. (2016) Calculation of gamma-ray attenuation parameters for locally developed shielding material: Polyboron. From * The Department of Nuclear Engineering, University of Dhaka, Dhaka 1000, Bangladesh, and ** The Bangladesh Atomic Energy Commission, Bangladesh. Journal of Radiation Research and Applied Sciences 9, (26–34).

3. Caldirola, Piero. (1953) Nuovo Cimento 10 1747; Suppl. Nuovo Cim. 3 (1956) 297; Rivista Nuovo Cim. 2 (1979), issue no.13, and refs.

4. Caldirola, Piero. (1980) The introduction of the chronon in the electron theory and a charged lepton mass formula. Lett. Nuovo Cim. 27, pp. 225-228.

5. Cartan, Élie. (1913) Projective groups that leave no multiplicity plane invariant, Bulletin of the S.M.F. Volume 41, Pages 53-96.

6. Cartan, Élie. (1923-1925) On manifolds with affine connection and the General Theory of Relativity (First Part), Ann. Sci. École Norm. Sup., 40 (1923) Pages 325-412, 41 (1924) Pages 1-25, and 42 (1925) Pages 17-88.

7. Dowling, M.R. and Nielsen, M.A. (November, 2008), The geometry of quantum computation, Quantum Info. Comput. 8, Pages (861, 899).

8. Everitt, C.W.F.; Muhlfelder, B.; DeBra, D.B.; Parkinson, B.W.; Turneaure, J.P.; Silbergleit, A.S.; Acworth, E.B.; Adams, M.; Adler, R.; Bencze, W.J.; et al. (2015) The Gravity Probe B test of General Relativity. Class. Quantum Gravity, 32, 224001.

9. Gao, Ping, Jafferis, Daniel Louis, and Wall, Aron C. (2019) Traversable Wormholes via a Double Trace Deformation, Center for the Fundamental Laws of Nature, Harvard University, Cambridge, MA, School of Natural Sciences, Institute for Advanced Study, Princeton, NJ.

10. Ginzburg, V. L. (July 2004) On superconductivity and superfluidity, what I have and have not managed to do, as well as on the physical minimum at the beginning of the 21st century. ChemPhysChem. 5 (7): 930–945.

11. Hawking, Stephen and Penrose, Roger. (1996) The Nature of Space and Time. Princeton University Press, Princeton, New Jersey. Isaac Newton Institute Series of Lectures. ISBN-13: 978-0691037912.

12. Hulse, R.A.; Taylor, J.H. (1975) Discovery of a pulsar in a binary system. Astrophysics. J., 195, L51–L53.

13. Landau, Lev D. and Lifschitz, Evgeny M. (1984) Electrodynamics of Continuous Media. Course of Theoretical Physics. Vol. 8. Oxford: Butterworth-Heinemann. ISBN 978-0-7506-2634-7.

14. Lewin, Walter; Van Der Klis, Michiel (2006), Compact Stellar X-ray Sources, Cambridge University Press, p. 159, ISBN 0-521-82659-4.

15. Maldacena, Juan, Shenker, Stephen H., Stanford, Douglas. (2015) A Bound on Chaos, School of Natural Sciences, Institute for Advanced Study Princeton, NJ, Stanford Institute for Theoretical Physics and Department of Physics, Stanford University Stanford, CA.

16. New American Standard Bible. (1995) NASB1995. Colossians 3:3, Cephas, AD 64–68.

17. Nieves, Robert. (2020) A Dynamic Theory of Space-Time: A Matter of Waves. Published by Kindle Direct Publishing, Amazon.com, Inc. ISBN 9798667276289.

18. Nieves, Robert. (2021) A Synthesis of Quantum Gravity. Published by Kindle Direct Publishing, Amazon.com, Inc. ISBN 9798715826565.

19. Oppenheimer, J.R. and Snyder, H. (1939) On Continued Gravitational Contraction. Phys. Rev. 56, 455 – Published September 1, 1939.

20. Oppenheimer, J.R. and G. M. Volkoff. (1939) On Massive Neutron Cores, Physical Review, vol. 55, no. 4, p. 374.

21. Page, D. N. (2013) Time Dependence of Hawking Radiation Entropy, JCAP 1309, 028.

22. Penrose, Roger. (1963) Asymptotic properties of fields and space-times. Physical Review Letters. 10 (2).

23. Poincaré, H. (1890) "Sur le problème des trois corps et les équations de la dynamique". Acta Math. 13: 1–270.

24. Poincaré, H. (1890) Œuvres VII, 262–490 (theorem 1 section 8).

25. Psaltis, D. (2008) Probes and tests of strong-field gravity with observations in the electromagnetic spectrum. Living Rev. Relativity, 11, 9.

26. Rindler, W., (1969) Essential Relativity: Special, General, and Cosmological, Van Nostrand, New York.

27. Shannon, Claude E. (July, 1948). A Mathematical Theory of Communication. Bell System Technical Journal. 27 (3): 379–423.

28. Smithells, C.J. (1997) Metals Reference Book, Vol III, Butterworths, London, 1967, 737ff.

29. Straumann, Norbert. (2019) SPG Mitteilungen Communications de la SSP Auszug-Extrait Milestones in Physics (17) On the evolution of Black Hole Physics. Physik-Institut, Universität Zürich.

30. Tolman, R.C. (1934) Effect of inhomogeneity on cosmological models, Proceedings of the national academy of sciences of the United States of America, vol. 20, no. 3, p. 169.

31. Tolman, R.C. (1939) Static solutions of Einstein's field equations for spheres of fluid, Physical Review, vol. 55, no. 4, p. 364.

32. U.S. Naval Observatory, The Astronomical Almanac, (2010) X-Ray Sources, gives a range of 235–1320 µJy at energies of 2–10 KeV, where a Jansky (Jy) is 10^{-26} Wm^{-2} Hz^{-1}.

33. Wikipedia. (2022) Adapted from https://en.wikipedia.org/wiki/Very-high-energy_gamma_ray.

34. Weyl, Hermann. (1923) Space, Time, Matter: Lectures on the General Theory of Relativity, Springer Berlin Heidelberg.

35. Will, C.M. (2015) Focus Issue: Gravity Probe B. Class. Quantum Gravity, 32, 220301.

36. Wilson, Kenneth G. The Renormalization Group and Critical Phenomena, Nobel lecture, 8 December 1982, Laboratory of Nuclear Studies, Cornell University, Ithaca, New York 14853.

37. Wolf, Bernhard. (1995) Handbook of ion sources. CRC Press. ISBN 0849325021, 9780849325021, pages 27 and 11.

38. Woodhead Publishing Series in Biomaterials. (2013) Standardisation in Cell and Tissue Engineering: Methods and Protocols, Edited by Vehid Salid, 80 High Street, Sawston, Cambridge CB22, 3HJ, UK.

www.ingramcontent.com/pod-product-compliance
Lightning Source LLC
Chambersburg PA
CBHW071356210526
45465CB00001B/115